Industrial Cloud-Based Cyber-Physical Systems

Armando W. Colombo · Thomas Bangemann
Stamatis Karnouskos · Jerker Delsing
Petr Stluka · Robert Harrison
François Jammes · Jose L. Martinez Lastra
Editors

Industrial Cloud-Based Cyber-Physical Systems

The IMC-AESOP Approach

 Springer

Editors
Armando W. Colombo
Schneider Electric
Marktheidenfeld
Germany

and

University of Applied Sciences Emden/Leer
Emden
Germany

Thomas Bangemann
Institut für Automation und Kommunikation
Magdeburg
Germany

Stamatis Karnouskos
Corporate Research
SAP
Karlsruhe
Germany

Jerker Delsing
Department of Systemteknik
Luleå University of Technology
Luleå
Sweden

Petr Stluka
Honeywell ACS Labs
Prague
Czech Republic

Robert Harrison
University of Warwick
Coventry
UK

François Jammes
Schneider Electric
Grenoble
France

Jose L. Martinez Lastra
Tampere University of Technology
Tampere
Finland

ISBN 978-3-319-38265-4   ISBN 978-3-319-05624-1   (eBook)
DOI 10.1007/978-3-319-05624-1
Springer Cham Heidelberg New York Dordrecht London

Foreword I

An Advisor's Remarks on the IMC-AESOP Contributions

The work of the IMC-AESOP Consortium has been a valuable addition to our understanding of the opportunities and complexities of automation with cloud connected systems. This book summarizes some of the major contributions of these experts and shows clearly the vision of the future regarding automation.

As exhibited by the works of the IMC-AESOP Consortium the Internet of Things, connected devices of all kinds continues to grow daily. This connectivity includes smart client devices from PCs to smart phones to control systems, cloud services, and even vehicles. These devices have sensors, such as motion, location/GPS, cameras, etc., in addition to the computation and connection capabilities. This coupled with powerful cloud servers opens an environment of opportunity for service based automation. With control services from the factory to the office to the home.

Managing the services has shown many challenges. There are billions of devices many with unique service control protocols that need to be considered. Some devices are new, SCADA ready, others old with unique schemes. To build a service layer the system must overcome these unique designs in a logical way and IMC-AESOP has shown the first step here. Clearly, additional standards to interfaces can help ease this complexity. Once you connect devices and can query them and control them it gives great opportunity for service designs. The IMC-AESOP Consortium has shown several including managing lubrication in a plan and suitably climate controlling a house in cold climate to optimize the balance between owner comfort and utility utilization. One can envision many applications in the service over a cloud area from factory, to office, to home. However, having this level of control requires exceptional security controls. This kind of control used maliciously can ruin a factory or a home quite easily if misused. Suitable authentication and identity services must be established to open these controls to the cloud.

Another challenge is the rapid changing of technologies and the associated necessity for their agile adoption, as the use cases and prototype implementations described in the following chapters are showing. Taking as an example the District Heating Application described in Chap. 10, when the project started in 2010, few cars "talked to the Internet" so the home climate control system used road sensors to detect when the automobile was on its way to the garage; a few years later today's automobiles have integrated GPS sensors and can communicate to the cloud. This is putting under question the necessity for keeping or eliminate the need for the road sensor, since the car itself could inform the home climate system "I am coming home, turn up the heat!"

Admittedly, this is an amazing view into the future of automation for many markets. However, to realize this vision the many challenges need to be addressed. This includes standards for services and control interfaces, industry adoption of the services and their standards. The business model must make sense for all parties to suitably engage. Finally, the security matters must be built in to the system since there is potential high risk to any user factory or consumer. I am sure that follow-on efforts can address these challenges in detail.

Finally, I would like to commend the IMC-AESOP effort for its contributions to the IEEE Industrial Electronics Society (IES). The project has shared many results with the IES community through its conferences and publications. The IMC-AESOP has been a great sponsor of the IES Industry Forum that engages discussion between industry and researchers within selected IES conferences. Their contributions to IES have been a noteworthy utilizing of the IEEE as a cooperative between industry and research to achieve a key useful goal.

Santa Clara, December 2013   Michael W. Condry, Ph.D.
                                                       CTO, Global Ecosystem Development
                                                                      Intel Corporation

Foreword II

Key Aspects of the European Strategic Research and Innovation Agenda

Providing a Service-Oriented Architecture approach for monitoring and control of Process Control applications is among the main objectives of IMC-AESOP. By achieving this, IMC-AESOP will help Industry to respond to its needs with an SOA that pushes further the state of art and to develop solutions for systems that are getting further in complexity and in heterogeneity. As an industry involved in the business of monitoring and exploiting critical infrastructure such as the SCADA business, Thales is expecting IMC-AESOP to address a major issue of providing an architecture that connects heterogeneous sensors for an efficient control for supervising and monitoring large, complex, and secure infrastructure.

In preparing the ARTEMIS Strategic Research and Innovation agenda (ARTEMIS: European Technology Platform for Embedded Systems), we have also advocated the importance of "Service-oriented Embedded System solutions to support production management and operation, covering functions and components across the different levels of an ISA-95 compliant Production Enterprise Architecture, from the sensor/actuator, throughout SCADA, MES to the upper ERP and with open service integration to society energy infrastructure adhering to standards like ISO 61850 for Substation Automation." IMC-AESOP is going in this direction, and with it ability to provide web-services-based solutions it also should respond to one of the major issues which is the flexible and agile adaptation to market demands, individual customization, reduced commissioning, cost-effectiveness, to cite some of the features expected in the new SOA solutions towards "the Future Perfect Plant"—as envisioned in IMC-AESOP.

The "Future Perfect Plant" approach will enable monitoring and control of information flow in a cross-layer way where dynamic integration of the system components will improve adaptation to business evolving needs.

We also need pilots to facilitate the creation of new business in innovating eco-systems achieving a "European Dimension" and combining the R&D efforts across Europe. IMC-AESOP is among the precursors in this approach as it allows de-risking of solutions that could be deployed in complex infrastructures by linking all the layers from devices to enterprise systems. Its consortium is gathering major actors and competencies for this type of research to be achieved.

This book presents an excellent summary of results from the Research, Development and Innovation activities performed within IMC-AESOP and the visions and outlooks addressed by the major actors behind those results.

Paris, December 2013   Laila Gide
                                Director for Advanced Studies Europe, Thales

Foreword III

Intangible Questions that will Affect the New Generation of Engineers

Engineers who are about to start their professional career are faced with many questions, some of them very interesting and exciting and some of them possibly a bit worrying as the technological world they are about to enter has pressing questions in store. So they will have to find their way into the right use of technology. The outlook of the possibilities for the integration of mechanics, electronics, and software puts them in a unique historic position as the options for matching virtual and real worlds have never been before so close before.

In particular this means they will have the chance to develop mechanical architectures supported by software tools and control power that enable the user to achieve systems that were non-existent before. One may argue that this was always the case for engineers just about to enter their professional lives. In a way this is true, but on the other hand, the globalisation of factories, the international competition and their immediate company environment forces them to think beyond the "simple technical solution". Technology, and this is at the basis of their toolset, alone is not able to give answers to the underlying complex questions.

The Service-Oriented Architecture paradigm and associated technologies described in this book have interesting promises in terms of building systems that allow much shorter ramp-up times and reduction of investment for equipment because at least parts of it will be reusable, engineering tools that help the designer to create the production line in parallel to designing the new product, empower the system creator to draw on numerous hardware solutions that only need some little touch up, and are on stock, to name just a few.

But as mentioned above, these are technological questions for which some of the solutions are described in the following chapters of this book. The next generation of engineers and for this reason also plant managers will have to come to grips with underlying questions which will determine their midterm success, the welfare of their families and the societies around them, and their respective strategic decisions. It is on their shoulders to develop new models on how to employ their engineering results in the development of products and above anything else how to produce these products. What is the underlying paradigm for the factory of the future? Is it the autonomous entity somewhere out in the green which works fully automatic? Materials are delivered by autonomous vehicles, orders come in via Internet, and process design and customer-specific products are generated by means of audio computer intercommunication with the customer or the engineer.

Already today it becomes obvious that the means to generate welfare and additional value is not in line with such models. Humans are the determining factor. In a way they carry the product in their thinking. They have capacities that are different from what can be achieved with a machine. When we understand better that the machinery we develop with technology has to be rethought in terms of how it can be best used for the support of humans and when we begin to see that this is more than just to create nice robots that helps the "stupid worker at the belt" than we may get back on the track of personalities like Henry Ford who is quoted of having said, that he not only wants to create an effective factory but that he also wants to enable his employees to buy the product they make.

New control and automation options like Service-Oriented Architecture have the potential to provide the next generation of engineers to work for such solutions successfully. Like the needle of a compass their way of thinking has to be attracted for such a model. I hope they will succeed and the contributions of this book make a first step forward in this direction.

Esslingen, December 2013   Dr. Christoph Hanisch
                                                 Head of Future Technology, Festo

Preface

Future Industrial Infrastructures are expected to be complex System of Systems (SoS) that will empower a new generation of today hardly realizable, or too costly to do so, applications and services. New sophisticated enterprise-wide monitoring and control approaches will be possible due to the prevalence of Cyber-Physical Systems (CPS), which have made Machine-to-Machine (M2M) interactions a key competitive advantage and market differentiator. This will be possible due to several disruptive advances, as well as the cross-domain fertilization of concepts and the amalgamation of IT-driven approaches in the traditional industrial automation systems.

The Factory of the Future (FoF) will rely on a large ecosystem of systems where collaboration at large scale will take place. Additionally with the emergence of Cloud Computing, it is expected that Cyber-Physical Systems will harness its benefits, such as resource-flexibility, scalability, etc., and not only enhance their own functionality but also enable a much wider consumption of their own data and services. The result will be a highly dynamic flat information-driven infrastructure that will empower the rapid development of better and more efficient next generation industrial applications while in parallel satisfying the agility required by modern enterprises.

Designing and operating the factory of the future means dealing with several challenges such as structural, operational, and managerial independence of the shop floor and enterprise constituent systems, interoperability, plug and play, self-adaptation, reliability, energy-awareness, high-level cross-layer integration and cooperation, event propagation and management, etc. The future "Perfect Agile Factory" will enable monitoring, processing, and control information flow in a cross-layer way. As such the different systems composing the whole enterprise will be part of a distributed ecosystem, where components, hardware and software, can dynamically be discovered, added or removed, and dynamically exchange information and collaborate. This cross-layer, intra-enterprise collaborative infrastructure will be driven by business needs exposed and managed as individual and/or composed services by the system's components.

The application of the Service-Oriented Architecture (SOA) paradigm to virtualise the shop-floor allows it to expose its capabilities and functionalities as

"Services". These "Services" can be located on physical resources, i.e., smart devices and systems, but also on the cyber-space identified here as "Shop Floor Service Cloud". This book introduces the vision and describe the major results of research, development and innovation works carried out by several major industrial players, leading universities and research institutes, within the European Collaborative Project "ArchitecturE for Service-Oriented Process—Monitoring and Control" (IMC-AESOP). More specifically, IMC-AESOP consists of a Research, Development and Innovation (R&D&I) approach that covers several aspects of the fusion of "Cyber-Physical Systems" and the "Service-Oriented Architecture and Cloud Computing," which tackled from the architecture, technology, migration, and engineering angles, and demonstrated through some selected industrial use-cases.

Going through the following pages, the reader will get a deeper view of the IMC-AESOP approach from multiple angles:

- Chapter 1 is dedicated being an introduction to what the reader can expect being presented within this book. It provides an overview of the motivation, vision, and efforts carried out by the partners of the IMC-AESOP project, toward defining the vision of cloud-based industrial CPS and demonstrating its advantages.
- Chapter 2 provides a short summary about today's situation and trends in automation. To be successful and take the potential user from where it is today, every innovation has to start from the latest state-of-the-art systems within the respective domain. While investigating the introduction of Service-Oriented Architectures to automation, and even down to the shop floor, latest standards, proofed technologies, industrial solutions, and latest research works in the automation domain have to be considered.
- Chapter 3 deals with bold architecture vision for cloud-based industrial systems. Future factories will rely on multi-system interactions and collaborative cross-layer management and automation approaches. Within this chapter a Service-Oriented Architecture is proposed attempting to cover the basic needs for the next generation SCADA/DCS systems, i.e., monitoring, management, data handling, and integration, etc., by taking into consideration the disruptive technologies and concepts that could empower future industrial systems.
- Chapter 4 focuses on assessment of promising technologies available in an industrial context, utilizing Service-Oriented Architecture-based distributed large scale Process Monitoring and Control. Aspects of integration, real-time support, distribution, event-based interaction, service-enablement, etc., are approached from different angles.
- Chapter 5 focuses on the step-wise introduction of Service-Oriented Architecture into process monitoring and control, which requires a systematic approach to migrate from legacy systems into the next generation SOA-based SCADA/ DCS systems. The migration procedure proposed aims to preserve the functional integration, organize the SOA cloud through grouping of devices, and maintain

the performance aspects such as real-time control throughout the whole migration procedure.

- Chapter 6 deals with engineering methods and tools. These are seen as key enablers for efficiently designing, testing, deploying, and operating any industrial automation infrastructures. An overview of the user and business requirements for engineering tools, including system development, modeling, visualization, commissioning, and change in an SOA engineering environment is provided. An appraisal of existing engineering methods and tools, appropriate to four IMC-AESOP industrial use case is presented, followed by the description of a tool cartography adequate for engineering systems based on the IMC-AESOP approach.

To better depict the advancements achieved with the application of the IMC-AESOP approach, a series of four chapters are dedicated to present field trials demonstrating and evaluating results of the IMC-AESOP investigations.

- Chapter 7 Here we explore how the Service-Oriented Architecture can ease the installation and maintenance of one of the lubrication system of the world largest underground iron mine run by LKAB in north Sweden, with a focus on migration aspects.
- Chapter 8 Here the high demand in scalability of SOA-based systems and the exposition of services on and consume services from the automation cloud is investigated.
- Chapter 9 Here it is shown that the energy management can benefit from the advantages of service orientation, event-driven processing and information models for increased performance, easier configuration, dynamic synchronisation and long-term maintenance of complicated multi-layer industrial process solutions.
- Chapter 10 Here it is illustrated how the IMC-AESOP approach support building System of Systems. Applied to the district heating domain, this chapter presents the major features associated to a smart house demonstration where six different, heterogeneous, distributed systems have been integrated.

Finally, in Chap. 11 it is argued that if the vision of future cloud-based industrial cyber-physical system infrastructures is to become a reality and broadly adopted, industrial consensus has to be built on the adoption of adequate technologies, methods, and tools. This means a considerable amount of technological, application-oriented, and human-oriented challenges has to be tackled. This chapter identifies some of these challenges that need to be addressed by future research, development, and innovation activities.

We hope you enjoy this book which will inspire you to further advance the bold vision presented here, so that one day in the near future it may represent an industrial reality.

Emden, Marktheidenfeld	Armando W. Colombo
Magdeburg	Thomas Bangemann
Karlsruhe	Stamatis Karnouskos
Luleå	Jerker Delsing
Prague	Petr Stluka
Warwick	Robert Harrison
Grenoble	François Jammes
Tampere	Jose L. Martinez Lastra

Disclaimer

The information and views set out in this publication are solely those of the author(s) and do not necessarily reflect the official opinion of their associated affiliation. Neither the companies, institutions, and bodies nor any person acting on their behalf may be held responsible for the use which may be made of the information contained therein. We explicitly note that this report may contain errors, inaccuracies, or errors or omissions with respect to the materials.

Acknowledgments

The authors would like to thank for their support the European Commission, and all the partners of the EU FP7 project IMC-AESOP (http://www.imc-aesop.eu). The IMC-AESOP book in your hands has been possible due to the direct or indirect work of several people who contributed fruitful ideas, discussions, experiments, guidance, etc., and we would like to acknowledge them here (alphabetic order).

Fredrik Arrigucci	Midroc, Sweden
Thomas Bangemann	ifak, Germany
Roberto Camp	FluidHouse, Finland
Oscar Carlsson	Midroc, Sweden
Farid Cerbah	Dassault Aviation, France
Armando Walter Colombo	Schneider Electric, University of Applied Sciences Emden/Leer, Germany
Michael Condry	Intel, USA
Jerker Delsing	Luleå University of Technology, Sweden
Paul Drews	APS Mechatronic, Germany
Jens Eliasson	Luleå University of Technology, Sweden
Charbel El Kaed	Schneider Electric, France
Laila Gide	Thales, France
Per Goncalves Da Silva	SAP, Germany
Mario Graf	SAP, Germany
Daniel Hahn	APS Mechatronic, Germany
Robert Harrison	University of Warwick, UK
Vladimir Havlena	Honeywell, Czech Republic
Christian Hübner	ifak, Germany
Ji Hu	SAP, Germany
Dejan Ilic	SAP, Germany
François Jammes	Schneider Electric, France
Eva Jerhotova	Honeywell, Czech Republic
Otto Karhumaki	FluidHouse, Finland
Stamatis Karnouskos	SAP, Germany

Petr Kodet	Honeywell, Czech Republic
Rumen Kyusakov	Luleå University of Technology, Sweden
Karri Lehmusvaara	Tampere University of Technology, Finland
Per Lindgren	Luleå University of Technology, Sweden
Jose L. Martinez Lastra	Tampere University of Technology, Finland
Andrei Lobov	Tampere University of Technology, Finland
Keijo Manninen	Honeywell, Finland
Stuart McLeod	University of Warwick, UK
Marco Mendes	Schneider Electric, Germany
Johannes Minor	Tampere University of Technology, Finland
Jesper Moberg	Midroc, Sweden
Kevin Nagorny	University of Applied Sciences Emden/Leer, Germany
Philippe Nappey	Schneider Electric, France
Johan Nessaether	Midroc, Sweden
Matthias Riedl	ifak, Germany
Rolf Riemenschneider	European Commission, Belgium
Keijo Ruonamaa	FluidHouse, Finland
Marek Sikora	Honeywell, Prague, Czech Republic
Jarkko Soikkeli	Prodatec Oy, Finland
Franz-Josef Stewing	Materna, Germany
Petr Stluka	Honeywell, Prague, Czech Republic
Nico Suchold	ifak, Germany
Giacomo Tavola	Politecnico Di Milano, Italy
Marcel Tilly	Microsoft, Germany
Pavel Trnka	Honeywell, Czech Republic
Marko Vainio	FluidHouse, Finland
Jeffrey Wermann	University of Applied Sciences Emden/Leer, Germany

Contents

Acronyms

AE	Alarms and Events
APC	Advanced Process Control
API	Application Programming Interface
APL	Active Production List
BPEL	Business Process Execution Language
BPEL4WS	Business Process Execution Language for Web Services
BPMN	Business Process Modelling Notation
CBM	Condition Based Maintenance
CDL	Choreography Description Language
CEP	Complex Event Processing
COTS	Commercial off-the Shelf
CPS	Cyber-Physical Systems
DCS	Distributed Control System
DDE	Dynamic Data Exchange
DPWS	Devices Profile for Web Services
EAM	Enterprise Asset Management
EDD	Electronic Device Description
EDDL	Electronic Device Description Language
EPR	End-Point Reference
ERP	Enterprise Resource Planning
ESP	Event Stream Processing
FDI	Field Device Integration
FDT	Field Device Tool
FPL	Finished Production List
GUI	Graphical User Interface
HCF	HART Communication Foundation
HDA	Historical Data Access
HMI	Human–Machine Interface
HSE	High Speed Ethernet
HTTP	Hypertext Transfer Protocol
I/O	Input/Output
ICT	Information and Communication Technologies
IMC-AESOP	ArchitecturE for Service-Oriented Process—Monitoring and Control

IoT	Internet of Things
IP	Internet Protocol
IT	Information Technologies
KPI	Key Performance Indicator
M&C	Monitoring and Control
MAS	Multi Agent System
MES	Manufacturing Execution System
MPC	Model Predictive Control
OEE	Overall Equipment Effectiveness
OLE	Object Linking and Embedding
OPC	Open Connectivity via Open Standards, Previously OLE for Process Control
PaaS	Platform as a Service
PET	Process Engineering Tools
PIMS	Process Information Management Systems
PLC	Programmable Logical Controller
QoS	Quality of Service
RAM	Random Access Memory
RTU	Remote Terminal Units
SCA	Service Component Architecture
SCADA	Supervisory Control and Data Acquisition
SNIR	Signal to Noise and Interference Ratio
SOA	Service-Oriented Architecture
SOAP	Simple Object Access Protocol
SoS	System of Systems
TCP/IP	Transmission Control Protocol/Internet Protocol
UA	Unified Architecture
UC	Use Case
UDDI	Universal Description Discovery and Integration
UDP	User Datagram Protocol
UI	User Interface
UML	Unified Markup Language
WS	Web Service
WSAN	Wireless Sensor Actuator Network
WSDL	Web Services Description Language
WSDM	Web Services Distributed Management
WSN	Wireless Sensor Network
XML	Extensible Markup Language

Chapter 1
Towards the Next Generation of Industrial Cyber-Physical Systems

Armando W. Colombo, Stamatis Karnouskos and Thomas Bangemann

Abstract Intelligent networked embedded systems and technologies, ranging from components and software to Cyber-Physical Systems (CPS) [1], are of increasing importance to the ICT supply industry, system integrators and all major mainstream sectors of the economy [9]. Development of new technologies for provisioning innovative services and products can lead to new business opportunities for the industry. Monitoring and Control are seen as key for achieving visions in several CPS dominated areas such as industrial automation systems, automotive electronics, telecommunication equipments, smart-grid, building controls, digitally driven smart cities, home automation, greener transport, water and wastewater management, medical and health infrastructures, online public services and many others [10]. This chapter introduces cloud-based industrial CPS and describes the first results of making it a reality for the Next Generation SOA-based SCADA/DCS systems. The reader can learn about the research, development and innovation work carried out by a set of experts collaborating under the umbrella of the IMC-AESOP project, for specifying, developing, implementing and demonstrating the major features of Intelligent Monitoring and Control Systems and the advantages of implementing them in different industrial process control environments.

A. W. Colombo (✉)
Schneider Electric, Marktheidenfeld, Germany
e-mail: armando.colombo@schneider-electric.com

A. W. Colombo
University of Applied Sciences Emden/Leer, Emden, Germany
e-mail: awcolombo@technik-emden.de

S. Karnouskos
SAP, Karlsruhe, Germany
e-mail: stamatis.karnouskos@sap.com

T. Bangemann
ifak, Magdeburg, Germany
e-mail: thomas.bangemann@ifak.eu

A. W. Colombo et al. (eds.), *Industrial Cloud-Based Cyber-Physical Systems,*
DOI: 10.1007/978-3-319-05624-1_1, © Springer International Publishing Switzerland 2014

1.1 Current Paradigms and Technologies Associated to CPS

Advances in computation and communication resources have given rise to a new generation of high-performance, low-power electronic components that have increased communication capabilities and processing power. This has led to new possibilities that enable improved integration of heterogeneous devices and systems, with particular emphasis on platform independence, real-time requirements, robustness, security and stability of solutions, among other major requirements.

Industrialists, researchers and practitioners are associating these advances with a 4th Industrial Revolution (referred to as Industrie 4.0 in Germany [15]) happening today, where physical 'Things' get connected to the Internet [12] allowing the real touchable world to integrate part of the cyber-space. With these foundations in CPS [1] and IoT, a number of different system concepts and architectures (e.g. www.iot-a.eu) have become apparent in the broader context of cyber-physical systems [1, 4, 20] over the past couple of years such as collaborative systems [11], Service-Oriented Architectures (SOA) [3], networked cooperating embedded devices and systems [22], cloud computing [2], etc.

The umbrella paradigm underpinning novel collaborative systems is to consider the set of intelligent system units as a conglomerate of distributed, autonomous, intelligent, proactive, fault-tolerant and reusable units, which operate as a set of cooperating entities [6]. These entities are capable of working in a proactive manner, initiating collaborative actions and dynamically interacting with each other to achieve both local and global objectives along the three basic collaboration axes (as depicted in Fig. 1.1) associated to any application domain and related infrastructure, i.e. enterprise, supply-chain and life-cycle axes [11].

From the physical device control level up to the higher levels of the business process management system, as defined in ISA-95 (www.isa-95.com), from suppliers through the enterprise to the customer [5], and from design through operation to recycling phases of an engineering system life cycle, collaboration will be enabled if, on one hand, the involved systems act and react on their environment, sharing some principal commonalities and, on the other hand, have some different aspects that complement each other to form a coherent group of objects that cooperate with each other to interact with their environment [22].

As we are moving towards Smart Cyber-Physical Systems and the 'Internet of Things', millions of devices, not all time smart, are interconnected, providing and consuming information available on the network and are able to exchange capabilities collaborating in reaching common goals. As these devices need to interoperate both at cyber and physical levels, the service-oriented approach seems to be a promising solution, i.e. each device should offer its functionality as standard services, while in parallel it is possible to discover and invoke new functionality from other services on-demand. These technologies can be leveraged to build advanced functionality into smart cyber-physical systems, thus enabling to build ad hoc new distributed

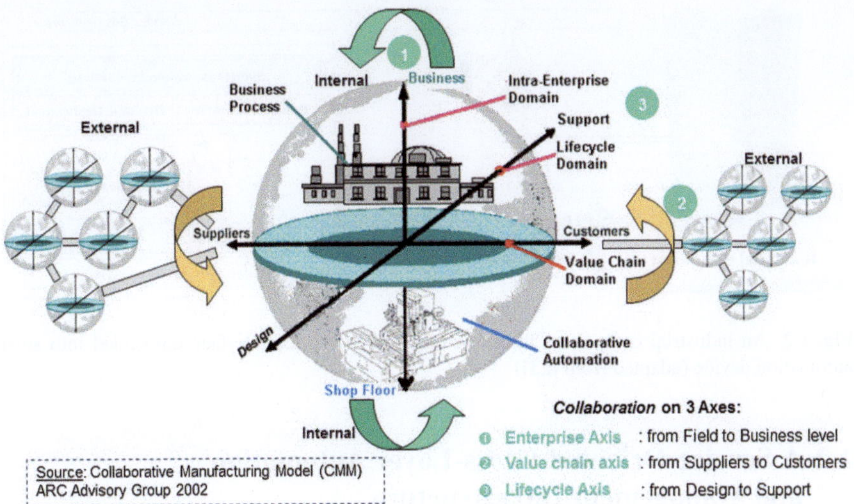

Fig. 1.1 Collaborative manufacturing model

application paradigms based on interconnected 'smart components' with a high level of autonomy.

This evolution towards global service-based infrastructures [6] indicates that new functionality will be introduced by combining services in a cross-layer form, i.e. services relying on the enterprise system, on the network itself and at device level will be combined. New integration scenarios can be applied by orchestrating the services in scenario-specific ways. In addition, sophisticated services can be created at any layer (even at device layer) taking into account and based only on the offered functionality of other entities that can be provided as a service [7, 18]. In parallel, dynamic discovery and peer-to-peer communication will allow to optimally exploit the functionality of a given device. It is clear that we are moving away from isolated stand-alone hardware and software solutions towards more cooperative models. However, in order to achieve that, several challenges need to be sufficiently tackled.

The convergence of solutions and products towards the SOA paradigm adopted for smart cyber-physical systems contributes to the improvement in the reactivity and performance of industrial processes such as manufacturing, logistics and others. This is leading to a situation where information is available in near-real-time based on asynchronous events, and to business-level applications that are able to use high-level information for various purposes, such as diagnostics, performance indicators, traceability, etc. SOA-based vertical integration will also help to reduce the cost and effort required to realise a given business scenario as it will not require any traditional high-cost solutions such as custom-developed device drivers or third-party integration solutions.

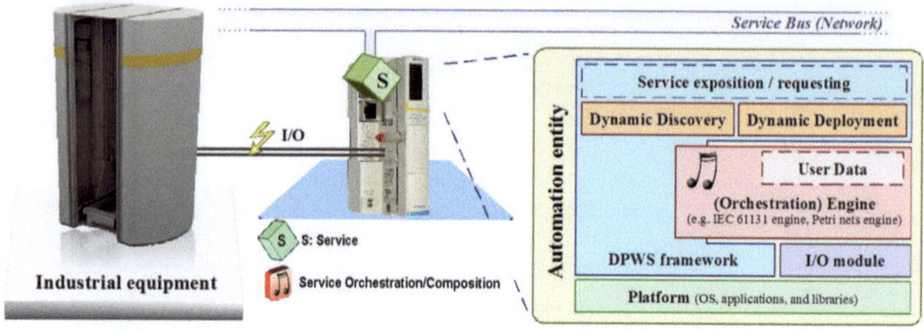

Fig. 1.2 An industrial component virtualised by a Web service interface embedded into smart automation device (adapted from [23])

1.2 A Service-Oriented Cross-Layer Automation and Management Infrastructure

A service-oriented cross-layer automation and management infrastructure adopts the 'collaborative automation' paradigm, combining cloud computing and Web services technologies [16], among others. The aim is to effectively develop tools and methods to achieve flexible, reconfigurable, scalable, interoperable network-enabled collaboration between decentralised and distributed cyber-physical systems. A first step towards this infrastructure is to create a service-oriented ecosystem. That is, networked systems composed of smart embedded devices that are Web service compliant (as depicted in Figs. 1.2 and 1.3), interacting with the physical and organisational environment and able to expose, consume and sometimes process (compose, orchestrate) services, pursuing well-defined system goals.

Taking the granularity of intelligence to the automation device level allows intelligent system behaviour to be obtained by composing configuration of devices that introduce incremental fractions of the required intelligence. From a runtime infrastructure viewpoint, the result is a new breed of flexible real-time embedded devices (wired/wireless) that are fault-tolerant, reconfigurable, safe and secure. Among other characteristics of such systems, automatic configuration management is a new challenge that is addressed through basic plug-and-play and plug-and-run mechanisms. The approach favours adaptability and rapid reconfigurability, as reprogramming of large monolithic systems is replaced by configuring loosely coupled embedded units.

The use of device-level service-oriented architecture contributes to the creation of an open, flexible and agile environment by extending the scope of the collaborative architecture approach through the application of a unique communication infrastructure [26], down from the lowest levels of the device hierarchy up into the manufacturing enterprise's higher level business process management systems [18]. The result of having a single unifying application-level communication technology across the enterprise, labelled as 'service bus (network)' in Fig. 1.4, transforms the

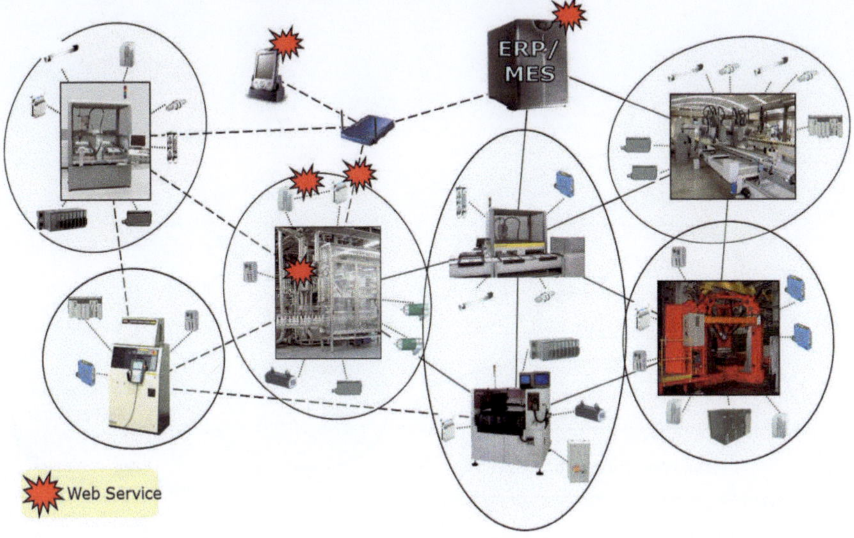

Fig. 1.3 An industrial system viewed as a distributed set of smart service-compliant devices and systems

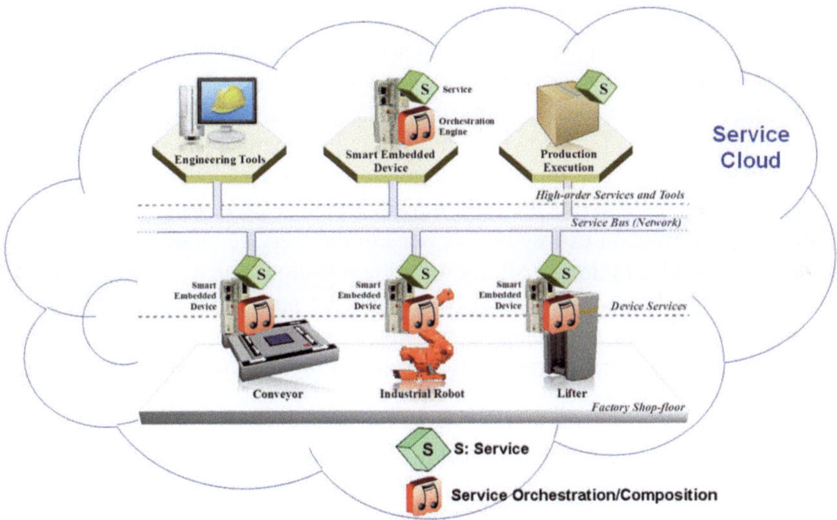

Fig. 1.4 A service-oriented view of an industrial system (adapted from [23])

traditional hierarchical view of the industrial environments into a flat automation, control and management infrastructure. That is, devices and systems located at different levels all have the same Web service interface and are able to interact. This functional interaction is completely independent of the physical location in the traditionally implemented enterprise hierarchy.

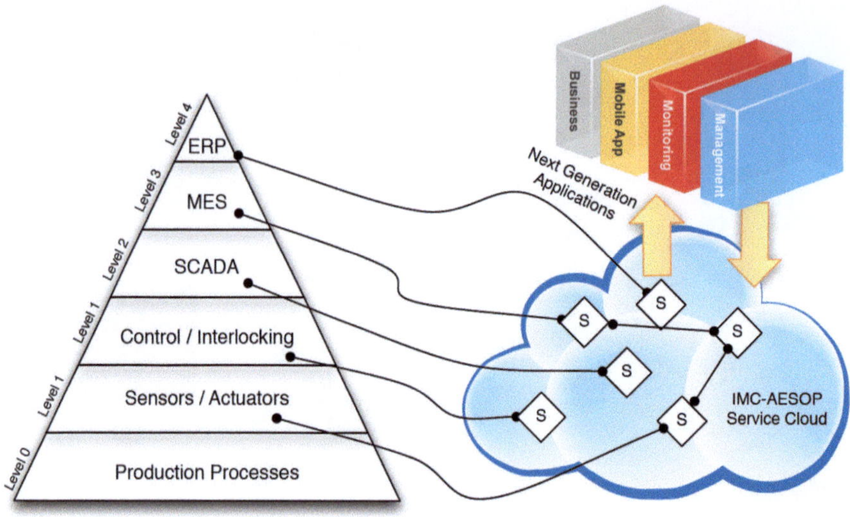

Fig. 1.5 Building supervisory control and management functions as applications using services exposed by devices and systems in the physical world and by the IMC-AESOP cloud in the cyber world

From a purely functional perspective, one of the major challenges is focussed, on one side, on managing the vastly increased number of intelligent devices and systems populating the collaborative SOA-based system and mastering the associated complexity. On the other side, following emerging requirements of control, automation, management and business applications, other challenges are engineering, development and implementation of the right infrastructure to make usable the explosion of available information exposed as services in the 'service cloud' originated, e.g. on the SOA-based shop floor [24]. Industrial applications can now be rapidly composed/orchestrated by selecting and combining the new services and capabilities offered as service in an automation cloud, which represents the partial or total virtualisation of the automation pyramid, as depicted in Fig. 1.5 and explained in Chap. 3 of this book.

1.3 Intelligent Service-Oriented Monitoring and Control: The IMC-AESOP Approach

The world market for technologies, products and applications alone that are related to what the Internet of Things enables, i.e. Monitoring and Control (M&C), will increase significantly in the next years. The world M&C market is expected to grow reaching 500€ billion in 2020. The M&C European market follows the same trends as that of the M&C world in terms of product repartition and market product evolution.

The European monitoring and control market will reach 143€ billion in 2020 [25]. When analysing the major application domains for real-time monitoring and control from the large process industry viewpoint, these indexes and the related expectations outline the tremendous potential and value.

Large process industry systems are a complex (potentially very large) set of (frequently) multidisciplinary, connected, heterogeneous systems that function as a complex distributed system whose overall properties are greater than the sum of its parts, i.e. very large-scale integrated devices (not all time smart) and systems whose components are themselves systems. Multidisciplinary in nature, they link many component systems of a wide variety of scales, from individual groups of sensors to whole control, monitoring, supervisory control systems, performing SCADA and DCS functions. The resulting combined systems are able to address problems which the individual components alone would be unable to do and to yield control and automation functionality that is only present as a result of the creation of new, 'emergent', information sources, and results of composition, aggregation of existing and emergent feature- and model-based monitoring indexes.

These very large-scale distributed process automation systems, that IMC-AESOP is addressing, constitute system of systems [14], and are required to meet a basic set of criteria known as Maier's criteria [21], i.e.:

1. Operational independence of the constituent systems
2. Managerial independence of the constituent systems
3. Geographical distribution of the constituent systems
4. Evolutionary development
5. Emergent behaviour.

Such systems should be based on process control algorithms, architectures and platforms that are scalable and modular (plug and play) and applicable across several sectors, going far beyond what current "Supervisory Control and Data Acquisition (SCADA)" systems, and Distributed Control Systems (DCS) and devices can deliver today.

A first fast analysis of current implemented SCADA and DCS systems detects a set of major hindrances for not completely fulfilling some of those criteria: the large number of incompatibilities among the systems, 'hard coded' data, different views on how systems should be configured and used, coexistence of technologies from very long periods of time (often more than 20 years), and use of reactive process automation components and systems instead of having them working in a proactive manner. If we began hooking all these hindrances, we would soon have an unmanageable mess of wiring and custom software, and little or no optimal communication. Today, this has been the usual result, where 'point solutions' have been implemented without an overall plan to integrate these devices into a meaningful 'Information Architecture'.

Looking at the latest reported R&D solutions for control and automation of large distributed systems, it is possible to identify today that there are already many known possibilities for covering some and, if possible, many or all the criteria addressed above. The IMC-AESOP concept points to optimisation at architectural and functional levels of the logical and physical network architectures behind process automation systems, mainly towards a potential optimal configuration and operation, e.g. of energy consumption [17] in current complex and power hungry process industries based on service-oriented process control algorithms, scalable and modular SOA-based Supervisory Control and Data Acquisition (SCADA) and distributed control systems (DCS) platforms, going far beyond what current centralised SCADA and DCS can deliver today [16].

To address integration of very large numbers of subsystems and devices, the IMC-AESOP approach takes its roots in previous work in several research and development projects [7, 13, 18], which demonstrated that embedding Web services at the device level and integrating these devices with MES and ERP systems at upper levels of an enterprise architecture was feasible not only at conceptual but also at industrial application level. The first results shown in pilot applications running in the car manufacturing, electromechanical assembly and continuous process scenarios have been successful, confirming that the use of cross-layer service-oriented architectures in the industrial automation domain is a promising approach, able to be extended to the domain of control and monitoring of batch and continuous processes.

Such an application domain, large process systems composed of very large numbers of systems, is challenging in terms of:

- Distributed monitoring and control of very large-scale systems (tens of thousands of interconnected devices are encountered in a single plant) enabling plant efficiency control, product and production quality control.
- A multitude of plant functions requesting information and functionality due to continuously changing and increasing business requirements.
- Integration of existing devices which generates the data and information necessary for multitude of plant functionalities like plant operation, maintenance, engineering, business and technology, i.e. system of systems integration, operation and evolution.
- The very large spread in device and system performance requirements regarding response time, power consumption, communication bandwidth and security.
- Legacy compatibility (20 year old systems have to interoperate with modern ones).

When using service-oriented architectures in process control applications, several advantages are expected. For open batch and/or process automation monitoring and control systems these include:

- The ability to be accessed by any other system of the enterprise architecture able to call other services.
- Improved ease-of-use and simplified operation and maintenance of SOA-based SCADA and DCS system embedded devices due to the universal integration capabilities that the services are offering.
- A next generation of SOA-based process automation components offering plug-and-play capabilities, providing self-discovery of all devices and services of the complete plant-wide system.

For proactive batch and/or process automation monitoring and control systems these include:

- The ability to expose their functionalities as services.
- The ability to compose, aggregate and/or orchestrate services exposed by themselves and from other devices in order to generate new distributed SCADA and DCS functions (also exposed as 'services' at the shop floor).
- At the shop floor these systems are interoperable with SOA-based systems of the upper levels of the enterprise architecture (e.g. integrating ERP and MES with the SCADA and DCS).
- A next generation of SOA-based devices and system exposing SCADA and DCS self-adaptable (emergent) functionalities (as a result of automatic service composition or orchestration), taking care of real-time changes in the dynamic system.
- The generation of new monitoring indexes and control functions at different levels of the plant-wide system, as a result of using event propagation, aggregation/orchestration/composition of services and management properties of the SOA-based distributed SCADA and DCS.

All the systems can benefit from cost-effectiveness, thanks to optimised SCADA and DCS distribution at the device level on the shop floor and at upper IT system levels. An additional benefit stems from the easier network management of large-scale networked systems. Based on these advantages a clear possibility is to generate system energy usage optimisation. With SOA-approach integration of subsystems having appropriate information, it can be done both at the operator and business levels, where different approaches to energy optimisation can be applied.

1.4 Positioning the IMC-AESOP Approach Within the Industrial Automation Landscape

The degree of reliability and efficiency of energy consumption/utilisation in the operation of industrial environments depends not only on the operation of the individual mechatronic/hardware components but also on the structure and behaviour of the embedded supervisory control system. Supervisory tasks have to be performed at two different and separate but networked levels, i.e. the shop floor and the upper levels of the enterprise architecture. At each of those levels it is possible to identify a set of functional and logical components that are responsible for performing the following functions: sensing, information collection, signal and information processing, decision-making and diagnosis and discrete-event control.

Each level (enumerated as 1–6 in Fig. 1.6) has its own time-constraints (from micro-seconds to days and weeks) and its own domain of data and information processing. Monitoring of operations, of the behaviour of the mechatronic/hardware components and of the system as a whole, is an essential function of such a supervisory control system. Consider the definition of 'monitoring' as the act of identifying the characteristic changes in a process and in the behaviour of mechatronic/hardware

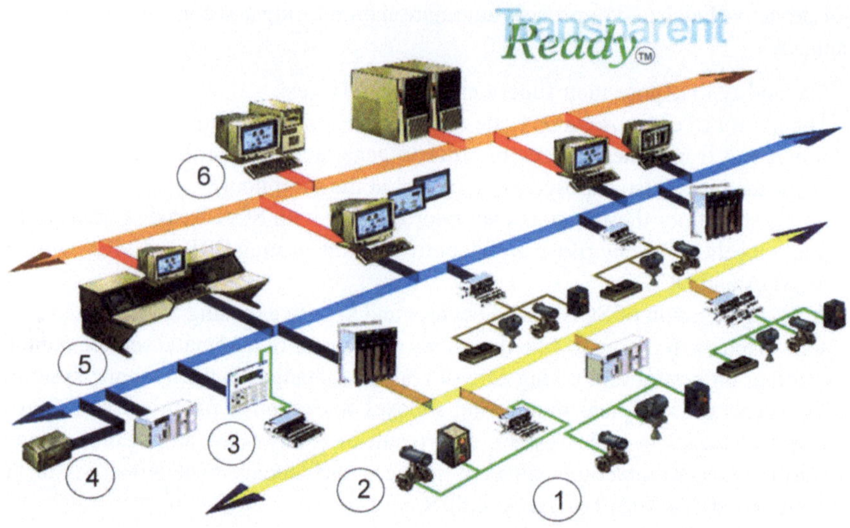

Fig. 1.6 Schneider electric enterprise system architecture 'Transparent Ready™'

resources by evaluating process and component signatures without interrupting normal operations [8].

In a plant, there are a set of process control stations that control different process sections in the plant, numbered 2 and 3 in Fig. 1.6. They are connected to various devices, distributed I/O stations, PLCs, etc., that are themselves connected to the process equipment labelled as number 1. For larger and process-specific equipment, the supplier also includes dedicated and unique devices, systems or complete control. In the overall plant monitoring and control, other systems and sections are also integrated like lubrication systems, transformers, switchgears, valves, ventilation, heating, etc. For operators, engineers, maintenance personnel and management, there are one or several control and engineering rooms available, as well as mobile devices for local monitoring and control, depicted as number 4 in Fig. 1.6. At the enterprise level, there are information access, control and analysis through various management and enterprise information and control systems, identified by numbers 5 and 6 in Fig. 1.6.

1.5 Introducing SOA and Cloud-Computing Paradigms into the Architecture of a Process Control System

IMC-AESOP proposed an infrastructure that goes well beyond existing approaches for monitoring and supervisory control, as depicted in Fig. 1.7. Following the development of computer network architectures, supervisory systems have undergone continuous evolution from a first generation based on centralised monolithic structure throughout, and second and third generations exploiting distribution and network-

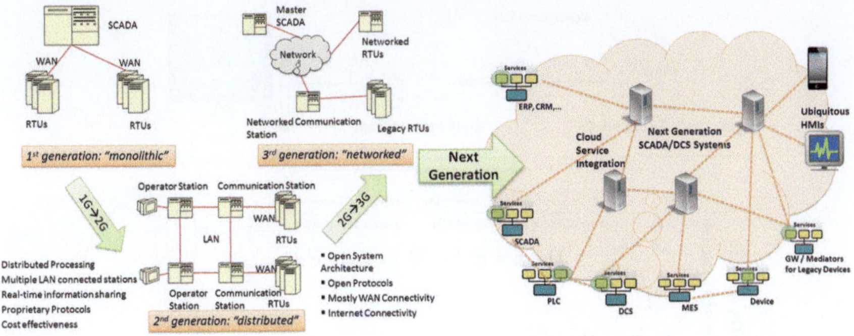

Fig. 1.7   IMC-AESOP impact on evolution of supervisory systems

ing capabilities. The next step was to evolve to a new service-oriented generation, called here 'The Next Generation SCADA/DCS systems', exposing functionalities and offering information that spans both domains, i.e. physical world and cyber world as represented in Fig. 1.7 by the service cloud.

This next generation of SCADA/DCS systems enable cross-layer service-oriented collaboration not only at horizontal level, e.g. among cooperating devices and systems, but also at vertical level between systems located at different levels of a Computer Integrated Manufacturing (CIM) or a Plant-Wide System (PWS) (http://www.pera.net). Focussing on collaboration and taking advantage of the capabilities of cooperating objects poses a challenging but also very promising change in the way future plants will operate, as well as in the way control and automation software will be designed. Also, the form to specify, model and implement the interactions among objects inside the plant. The future 'Perfect Plant' [6, 19] will be able to seamlessly collaborate and enable monitoring and control information flow in a cross-layer way. As such, different systems are part of a SCADA/DCS ecosystem, where components (devices and systems) can be dynamically added or removed, where data and information are exposed as services, where dynamic discovery enables the on-demand information acquisition and where control, automation and management functions can be performed as composition, orchestration, choreography of those services.

All current systems migrated under the SOA-based paradigm start being capable to share information in a timely and open manner, enabling an enterprise-wide system of systems that dynamically evolves based on business needs. With this approach, industrialists, researchers and practitioners also target future compliance and follow concepts and approaches that start enabling to design today the perfect 'legacy' system of tomorrow. That is, a system being able to be easily integrated in long-running infrastructures (e.g. in the chemical industry with a lifetime of 15–20 years).

The SOA-based approach, proposed by IMC-AESOP and explained in the following chapters, when applied to manufacturing and process control systems, allows on one hand, to present a set of SCADA and DCS functionalities as services, simplifying in this manner the integration of monitoring and control systems on application layer. On the other hand, the networking technologies that are already known to control

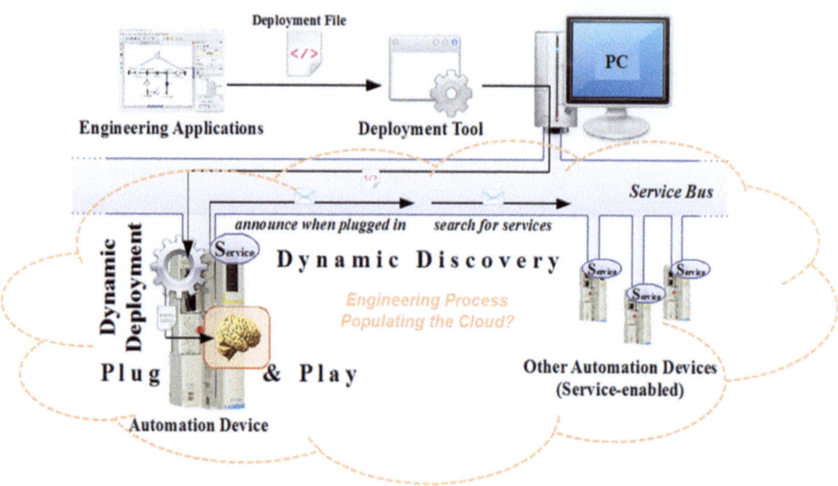

Fig. 1.8 Challenging the engineering—populating the automation cloud

engineers, could also simplify the inclusion of, or migration from, existing solutions into the next generation SCADA and DCS systems at network layer.

To achieve this, the focus of the research, development and innovation works has been put onto collaborative large-scale dynamic systems combining physical devices and systems with cloud-based infrastructure. Architectures and platforms that are scalable and modular (plug and play) and are applicable across several sectors have been implemented supporting the cyber-physical infrastructure. Populating the cloud-based infrastructure with the adequate cyber (and physical systems) presents another set of challenges to engineers and specialists responsible for 'Engineering' the manufacturing and process control and automation systems as depicted in Fig. 1.8. Starting with the connectability of devices and systems, followed by interoperability that facilitates collaboration, a new form of component-functional-oriented thinking affects the development and use of the whole set of engineering methods and tools along the engineering life cycle.

To populate the cloud with the right and necessary services exposed in cyberspace by smart SOA-compliant devices and systems located in physical-space is the first obligatory step towards the realisation of the vision addressed above. However, clearly the vision goes far beyond what current SCADA and DCS can deliver. Collaborations will be able to be created dynamically, serve specific purposes and will span multiple domains, as explained later in Chap. 11.

To sum, the advent of the SOA paradigm for application in management and automation presents a significant aid to manufacturers in today's industrial challenges. The availability of SOA-ready smart devices and systems with associated or even built-in monitoring and other supervisory control services delivers to production engineers a new way of looking at the industrial environment. It is opening new

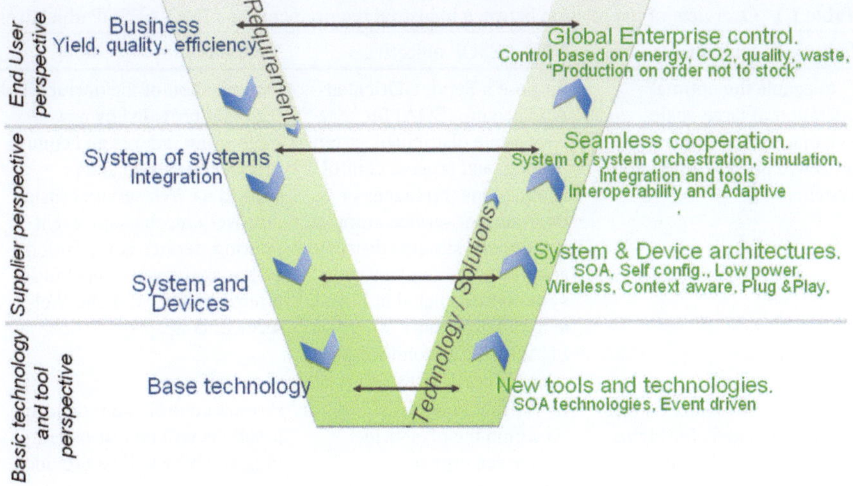

Fig. 1.9 State-of-the-art perspectives

avenues to visualise the evolution of systems and associated processes by making available a more detailed visualisation of the system's status in real-time.

1.6 The IMC-AESOP Approach: Beyond the State-of-the-Art

The IMC-AESOP approach builds on top of well-known scientific and technological trends such as virtualisation, software-as-a-service, cloud computing, collaborative automation, cooperative objects, etc., and responds to main industrial requirements as summarised in Fig. 1.9.

In the following sections, the progress beyond what is known today, reached applying the IMC-AESOP approach, is briefly described in three major dimensions: (i) end-user perspective, (ii) supplier perspective and (iii) tools and basic technology perspective. Table 1.1 depicts the relationship between these perspectives and the innovation aspects addressed by the approach. An intensive and carefully prepared analysis of the state of the art in industrial automation is presented in Chap. 2.

1.6.1 End-User Dimension

The industrial state of the art of large process control systems can be exemplified by the latest LKAB investment in their KK-4 pellets plant,[1] which was taken in production in early 2009. The system has more than 23.000 I/Os running in classi-

---

[1] http://www.lkab.com/en/Future/Investments/Refining/

Table 1.1 Overview of the relation between industrial requirements and IMC-AESOP objectives

Industrial requirements	IMC-AESOP objective	Thoughts
…to enable the optimal operation of large-scale dynamic systems through proactive process automation systems	Propose a Service-Oriented Architecture (SOA) for very large-scale distributed systems in batch and process control applications (up to tens of thousands of service-compliant devices and systems distributed across the whole plant-wide system (as depicted in Figs. 1.3 and 1.4) exposing SCADA/DCS monitoring and control functions as services	Optimisation of the operation of the plant provided by new monitoring indexes and control functions exposed and/or applied as Web services (using discovery mechanism, event filtering, service composition and/or aggregation capabilities offered by the SOA and Web services concepts)
Proactiveness requires novel predictive models for higher performance and fault adaptation and recovery. The architectures should enable QoS, and reduce the reconfiguration effort	Investigate how 'deep' we can go within the plant-wide system (enterprise architecture) with SOA-based monitoring and control models and functions (are we able to get SOA at the device level inside process control loops?)	Process control and monitoring functions will be distributed. Plug and play will be provided by discovery mechanisms, which will be extended to work for large-scale distributed systems
Proactiveness requires novel predictive models for higher performance and fault adaptation and recovery	Build a foundation for predictive performance of such service architecture based on a formal approach to event-based systems	Investigations will determine if event-based mechanism can be used for process control loops and if sufficient performance for use in the lowest levels of control loops can be achieved
Such systems should be based on architectures and platforms that are scalable and modular (plug and play) and are applicable across several sectors, going far beyond what current SCADA and DCS can deliver today	Investigate the co-habitat of currently used synchronous SCADA and DCS with the new asynchronous SOA-based monitoring and control system, going beyond what the current implemented control and monitoring systems are delivering today	It should be possible to build many different SCADA and DCS functions by combining the current centralised with the new SOA-based systems
The architectures should facilitate reuse	Propose a transition path from legacy systems (e.g. a 20 year-old machine) to an SOA-compliant system. To investigate how today's DCS structures (runtime as well as engineering) can be mapped to SOA, exploding the natural similarities that seem to exist	The transition path should consider the requirement that the new SOA-based process control system has to be an adequate legacy system in the next 5–10 years
…a new generation of open and proactive batch and process automation monitoring and control systems, and to address associated standardisation	Contributing to relevant standardisation bodies like IEC65E (IEC 61512-1 and -2), based on the former IS SP88, NAMUR NE33, OASIS (e.g. WS-DD WG) etc.	

cal hierarchical control architecture. Parallel to the control system, they have other systems, e.g. for maintenance.

End-users like LKAB run a number of such large process control systems, continuous or batch. They have already identified areas where cooperation between systems like those discussed above can generate large benefits regarding production efficiency, product quality control, energy usage optimisation and CO_2 minimisation.

Research projects like 'Mine of the future'[2] have been providing results targeting the needs for increased integration of ICT-based systems. Here, the capability of seamless and timely integration of data and information between systems and functionalities is identified as critical. These capabilities have to be flexible to handle continuously changing business and technologies.

Progress Beyond the State of the Art: Based on the SOA approach supported by standard-based and formal-based software design methods, the IMC-AESOP approach has been applied to define architectures (see Chap. 3), technologies (see Chap. 4), migration strategies (see Chap. 5) and methods and tools (see Chap. 6) suitable for addressing seamless and timely integration of data and information from SOA-compliant subsystems and devices. Altogether, this opens the door for larger improvements in the flexibility of monitoring and control of very large systems. Thus, it makes possible from the viewpoints of economics and manpower to address knowledge improvement possibilities regarding product and production quality as well as energy usage optimisation.

1.6.2 Supplier Dimension

The FP6 SOCRADES (www.socrades.eu) project evaluated several SOA solutions, applicable at the device level, including DPWS and OPC-UA, in the context of manufacturing automation. The DPWS solution was provided as a complete open-source software component, which was embedded in several devices and tools, and was successfully demonstrated in the car-manufacturing domain, in electromechanical flexible assembly systems, in continuous process control and in mechatronics interoperability trials. A potential merger between DPWS and OPC-UA was also investigated. Potential solutions were identified to reduce the costs of embedding DPWS in very simple devices. A first set of generic and automation-application services were identified, specified, developed and implemented in pilot industrial applications. Complementary to the results, the ITEA3 SODA project (https://itea3.org/project/SODA.html) looked at the ecosystem required to build, deploy and maintain an SOA application in several application domains (industrial, home, automotive, telecommunication, etc.).

However, none of these projects was addressing the specific challenging requirements coming with the engineering, development, implementation and operation of

---

[2] http://www.rocktechcentre.se/completed-projects/conceptual-study-smart-mine-of-the-future-smifu/

large-scale distributed systems for batch and continuous process applications. Major issues and associated challenges arise when SCADA/DCS functions have to be performed, e.g.:

- How deep into the system is it possible to go with SOA-based monitoring and control solutions (including associated costs, real-time and security, among other issues)?
- How can monitoring and control (SCADA) services with real-time aspects be modelled, analysed and implemented?
- How can a system be managed, when it is composed of thousands with SCADA-functionality (in the overall system, which may be composed of many different control loops, each one with several devices)?

Progress Beyond the State of the Art: The IMC-AESOP approach proposes and prototypically implements SOA-based components and systems for monitoring and control of very large industrial systems (see Chaps. 7, 8, 9 and 10). The technology-posed limits for SOA on subsystems and devices were investigated regarding real-time, event aggregation and filtering, event-driven mechanisms, etc.

1.6.3 Tools and Basic Technology Dimension

Currently, the tools and basic technologies supporting SOA for seamless and timely integration of data and information from subsystems and devices, and related communication systems, are based on standard programming languages like C and Java and operating systems like Linux, Windows and a variety of RTOSs.

Progress Beyond the State of the Art: Applying the IMC-AESOP approach means to investigate and introduce formal-based technologies, thus open for the automated validation of SOA-based system structural and behavioral specifications, e.g. orchestration topologies, the automated verification of code functionality guaranteeing real-time performance, making code generation, debugging and verification more economical (see Chap. 10).

1.7 Validation of the IMC-AESOP Approach in Selected Use Cases

The next generation of SCADA/DCS systems, as envisioned in IMC-AESOP, address the major needs of the end-users of distributed control systems within large-scale environments. The four use cases depicted in Fig. 1.10 express the wishes of end-users leading to technological improvements in monitoring and process control. These use cases listed hereafter span from an evolutionary process (starting from a process controlled in a classical way) with migration to an event-driven approach, to a complete system controlled and monitored based on the new approaches envisioned by

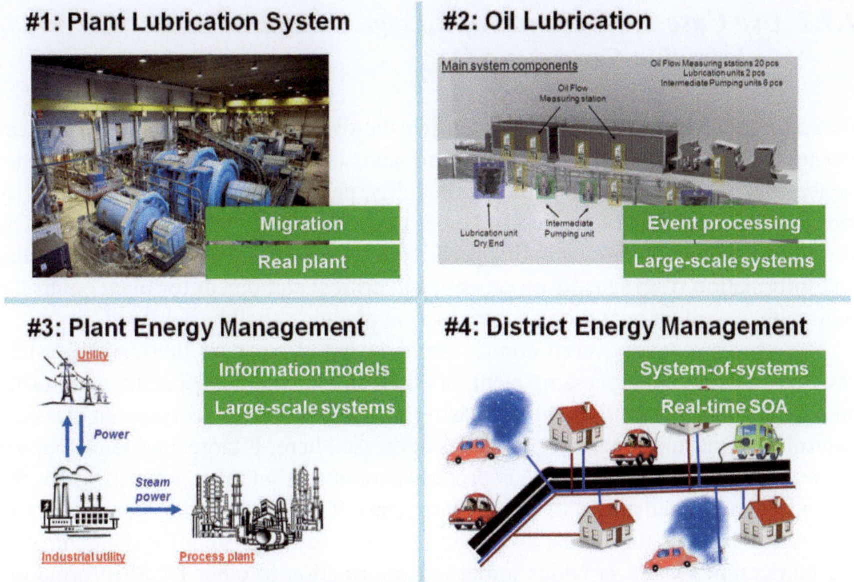

Fig. 1.10 Use cases validating the IMC-AESOP approach

IMC-AESOP, and even extends to combining systems to system-of-systems, specifically addressing monitoring, control and orchestration/composition of cross-domain services. The wide range of applications illustrates the needs to build new concepts applicable across several sectors.

Besides the individual use cases raised by individual end-users targeting specific applications, there are several aspects of common nature, applicable to the different use cases listed below.

These aspects are:

- *Isolation aspect with long time view*: A part of the plant has to be perpetual in order to be architecturally and functionally integrateable (this feature is called 'future compatible' with new coming 'emerging technology/components'). This can be realised through (i) description of a plant/plant-section as Web service-compliant (Architecture), where functions are exposed as Web services and (ii) SW implementation of a gateway or mediator.
- *Building a 'New Generation' plant* (*prototype or simulation*): A full plant or plant-section may be built with IMC-AESOP-compliant technology.
- *SCADA-DCS functions* (*aspects*) *as services*: Aspects will be specified and deployed into a pilot plant built as described above. This might be applicable to, e.g. Asset Management.

1.7.1 Use Case 1: Migration of a Legacy Plant Lubrication System to SOA

Industries continuously work on increasing the overall plant and equipment effectiveness, which leads to increasing requirements on open systems and much better system integration, availability, maintainability, performance, quality, functionality etc. The use-case scenario addressed here is targeting the plant control to increase the overall plant performance including predictive maintenance. With increased quality and information from sensors on process and critical equipment for plant control, a more effective plant operation and production planning shall be achieved.

The use case is an overall control scenario based on plant lubrication system installed in a mineral processing plant, or other similar suppliers-specific monitoring and control systems equipment addressing the migration aspects between classical control systems and the new approaches addressed here. It targets systems that are there for the function of numerous process equipments and that are critical for the operation and effectiveness of these equipments. The lubrication systems are typical critical systems for almost all process industries. The system for the control of the lubrication system provides important information to other DCS. Information that can be used by operators to avoid critical and damaging incidents, by operators, planning staff and management to improve production and plant efficiency and by the maintenance staff and management to analyse and improve the predictive maintenance.

The information provided is about the equipment itself and the consequences of malfunction but also on the sensor, system or infrastructure. The trail focusses on a system and equipment that must provide much better information in order to increase production availability and effectiveness and at the same time decrease work like daily maintenance. It shows integration of IMC-AESOP devices into a legacy system at an end-user, like a mineral processing plant at LKAB, Kiruna, Sweden. More detailed information is found in Chap. 7 of this book.

1.7.2 Use Case 2: Implementing Circulating Oil Lubrication Systems Based on the IMC-AESOP Architecture

The hydraulic control in industry is often used in applications where electrical drives cannot provide enough power. In fluid automation, the latest technologies could provide solutions that could allow better performance of the hydraulic systems. One important type of the processes found in fluid automation is the oil lubrication process, which is of demand in the pulp and paper, steel, and gas and oil industries, to name a few. Application of oil lubrication systems to large distributed systems brings new challenges such as strict environmental regulations. The new technologies can address this challenge by reducing the costs (both environmental and production) associated with oil exchange, thanks to advanced monitoring techniques of oil quality.

The oil lubrication systems found in paper machines could include dozens of lubricated nodes (gear boxes). The application of smart metres can make it possible to identify different parameters of the lubrication oil and make a conclusion on the need for maintenance work. Applying the IMC-AESOP approach, FluidHouse (www.fluidhouse.fi) achieved an increase in performance of large-scale distributed systems by

- Application of advanced measurement techniques;
- Information collection and processing with next generation SCADA systems based on standardised and widely accepted communication protocols.

It should be noted that the later item refers not to the old existing standards but to emerging IT standards and their applicability in industrial applications, e.g. SOA-related standards. More detailed information is found in Chap. 8 of this book.

1.7.3 Use Case 3: Plant Energy Management

A steam generation unit (steam boilers) provides steam for other units in the plant (process steam) and also drives turbo-generators. Generated electricity is used in the plant itself and/or is supplied to the power grid. In case of energy peaks, the plant may consume electricity from the grid. Steam generation consists of several boilers connected to a common header or to the system of common headers if more different levels of pressure are produced by the boilers. Overall, steam production may be split into independent sub-plants connected via steam transfer line.

Optimisation of such a system provides hierarchical overall plant optimisation across several levels: (i) base/device level, (ii) unit level, (iii) plant level and (iv) global level. Several basic requirements can be derived:

- *Model Consistency*: The basic and critical requirement for hierarchical optimisation is a consistency of model in all levels of optimisation. For instance, if a boiler has to be operated on a higher O_2 level due to problems with a mill, its efficiency drops. If the new efficiency curve is not propagated to higher levels, the benefits from optimal load allocation may be lost completely.
- *Integration issues*: Large-scale plants usually have some kind of optimisation controllers implemented on device level, but it can be difficult to get the right information in the right form to higher level optimisers.
- *Event-driven processing*: Some changes in a plant (e.g. boiler shutdown, closing of a transportation pipe) may require reconfiguration of optimisation problem in a remote optimiser. Such events must be communicated from a device to an optimiser.

- *'What-if' analyses*: Optimisers and all levels should support, in addition to real-time optimisation, on-demand optimisation for what-if analyses. On-demand optimisation should run the same algorithm as the real-time one, but on the data provided by a user. Usually, only a subset of data is provided by a user and the rest is taken from the process.

This use case evaluates how to address these requirements while using service-oriented approaches. More detailed information is found in Chap. 9 of this book.

1.7.4 Use Case 4: Building System of Systems with SOA Technology: A Smart House Use Case

The purpose of this use case is to investigate how a SCADA/DCS-like functionality can be generated by composing monitoring and control services. It primarily demonstrates how a domestic home and its supply and distribution systems for heat and electricity can be integrated to smart power grids and smart heat grids. This will be paired with the detection of incoming and outgoing vehicles exposing and consuming services inside an SOA-based transportation system. For this purpose, necessary services are specified, implemented and deployed on resource-constrained devices. It addresses aspects and characteristics associated to the system-of-systems paradigm, as the district monitoring scenario (system) covers heating, electricity and transportation systems. More detailed information is found in Chap. 10 of this book.

1.8 Conclusion

A number of new different system concepts and paradigms have become apparent in the broader context of cyber-physical systems [20] over the past couple of years such as collaborative systems [11], Service-Oriented Architectures (SOA) [3], networked cooperating embedded devices and systems [22], cloud computing [2], etc. This chapter presented the major aspects related to the vision of cloud-based industrial CPS. It is an introductory chapter briefing the research, development and innovation work carried out by a set of experts collaborating under the umbrella of the IMC-AESOP project, for specifying, developing, implementing and demonstrating major features of a next generation SOA-based SCADA/DCS systems and the advantages of implementing them in different industrial process control environments. The depicted IMC-AESOP efforts constitute a prelude to the CPS and Industry 4.0 vision [1, 15].

Acknowledgments The authors thank the European Commission for their support, and the partners of the EU FP7 project IMC-AESOP (www.imc-aesop.eu) for the fruitful discussions.

References

1. acatech (2011) Cyber-physical systems: driving force for innovation in mobility, health, energy and production. Technical report, acatech—National Academy of Science and Engineering. http://www.acatech.de/fileadmin/user_upload/Baumstruktur_nach_Website/Acatech/root/de/Publikationen/Stellungnahmen/acatech_POSITION_CPS_Englisch_WEB.pdf
2. Badger L, Grance T, Patt-Corner R, Voas J (2012) Cloud computing synopsis and recommendations. Technical report, NIST Special Publication 800-146, National Institute of Standards and Technology (NIST). http://csrc.nist.gov/publications/nistpubs/800-146/sp800-146.pdf
3. Boyd A, Noller D, Peters P, Salkeld D, Thomasma T, Gifford C, Pike S, Smith A (2008) SOA in manufacturing—guidebook. Technical report, IBM Corporation, MESA International and Capgemini. ftp://public.dhe.ibm.com/software/plm/pdif/MESA_SOAinManufacturingGuidebook.pdf
4. Broy M (2013) Cyber-physical systems: concepts, challenges and foundations. ARTEMIS magazine (14). http://www.artemis-ia.eu/publication/download/publication/877/file/ARTEMISIA_Magazine_14.pdf
5. Camarinha-Matos L, Afsarmanesh H (2008) Collaborative networks: reference modeling. Springer, New York
6. Colombo AW, Karnouskos S (2009) Towards the factory of the future: a service-oriented cross-layer infrastructure. In: ICT shaping the world: a scientific view. European Telecommunications Standards Institute (ETSI), Wiley, New York, pp 65–81
7. Colombo AW, Karnouskos S, Mendes JM (2010) Factory of the future: a service-oriented system of modular, dynamic reconfigurable and collaborative systems. In: Benyoucef L, Grabot B (eds) Artificial intelligence techniques for networked manufacturing enterprises management. Springer, London. ISBN 978-1-84996-118-9
8. Du R, Elbestawi MA, Wu SM (1995) Automated monitoring of manufacturing processes, Part 1: monitoring methods. J Eng Ind 117(2):121–132. http://dx.doi.org/10.1115/1.2803286
9. European Commission (2013a) Cyber-physical systems: uplifting Europe's innovation capacity. http://ec.europa.eu/digital-agenda/en/news/cyber-physical-systems-uplifting-europes-innovation-capacity
10. European Commission (2013b) ICT for societal challenges. Publications Office of the European Union—Luxembourg. doi:10.2759/4834, http://ec.europa.eu/information_society/newsroom/cf/dae/document.cfm?doc_id=1944
11. Harrison R, Colombo AW (2005) Collaborative automation from rigid coupling towards dynamic reconfigurable production systems. In: 16th IFAC world congress, vol 16. doi:10.3182/20050703-6-CZ-1902.01571
12. ITU (2005) ITU internet report 2005: the internet of things. Technical report, International Telecommunication Union (ITU)
13. Jammes F, Smit H (2005) Service-oriented architectures for devices—the sirena view. In: 3rd IEEE international conference on industrial informatics, INDIN '05, 2005, pp 140–147. doi:10.1109/INDIN.2005.1560366
14. Jamshidi M (ed) (2008) Systems of systems engineering: principles and applications. CRC Press, Boca Raton
15. Kagermann H, Wahlster W, Helbig J (2013) Recommendations for implementing the strategic initiative INDUSTRIE 4.0. Technical report, acatech—National Academy of Science and Engineering. http://www.acatech.de/fileadmin/user_upload/Baumstruktur_nach_Website/Acatech/root/de/Material_fuer_Sonderseiten/Industrie_4.0/Final_report__Industrie_4.0_accessible.pdf
16. Karnouskos S, Colombo AW (2011) Architecting the next generation of service-based SCADA/DCS system of systems. In: 37th annual conference of the IEEE industrial electronics society (IECON 2011), Melbourne, Australia
17. Karnouskos S, Colombo A, Lastra J, Popescu C (2009) Towards the energy efficient future factory. In: 7th IEEE international conference on industrial informatics, INDIN 2009, pp 367–371. doi:10.1109/INDIN.2009.5195832

18. Karnouskos S, Savio D, Spiess P, Guinard D, Trifa V, Baecker O (2010) Real world service interaction with enterprise systems in dynamic manufacturing environments. In: Artificial intelligence techniques for networked manufacturing enterprises management. Springer, London
19. Kennedy P, Bapat V, Kurchina P (2008) In pursuit of the perfect plant. Evolved Media, New York
20. Lee EA, Seshia SA (2011) Introduction to embedded systems: a cyber-physical systems approach, 1st edn. http://leeseshia.org
21. Maier MW (1998) Architecting principles for systems-of-systems. Syst Eng 1(4):267–284
22. Marrón PJ, Karnouskos S, Minder D, Ollero A (eds) (2011) The emerging domain of cooperating objects. Springer, Berlin. http://www.springer.com/engineering/signals/book/978-3-642-16945-8
23. Mendes JM (2011) Engineering framework for service-oriented industrial automation. Ph.D. thesis, Faculty of Engineering, University of Porto. http://paginas.fe.up.pt/niadr/NIADR/thesis_jmm_2011.03.21.pdf
24. Mendes JM, Bepperling A, Pinto J, Leitao P, Restivo F, Colombo AW (2009) Software methodologies for the engineering of service-oriented industrial automation: the continuum project. In: IEEE 37th annual computer software and applications conference 2013, vol 1, pp 452–459. http://doi.ieeecomputersociety.org/10.1109/COMPSAC.2009.66
25. DECISION (2008) Monitoring and control: today's market, its evolution till 2020 and the impact of ICT on these. European Commission DG Information Society and Media. http://www.decision.eu/smart/SMART_9Oct_v2.pdf
26. Ribeiro L, Barata J, Colombo A, Jammes F (2008) A generic communication interface for dpws-based web services. In: 6th IEEE international conference on industrial informatics, INDIN 2008, pp 762–767. doi:10.1109/INDIN.2008.4618204

Chapter 2
State of the Art in Industrial Automation

Thomas Bangemann, Stamatis Karnouskos, Roberto Camp, Oscar Carlsson, Matthias Riedl, Stuart McLeod, Robert Harrison, Armando W. Colombo and Petr Stluka

Abstract In the last decades, industrial automation has become a driving force in all production systems. Technologies and architectures have emerged alongside the growing organisational structures of production plants. Every innovation had to start

T. Bangemann (✉) · M. Riedl
ifak, Magdeburg, Germany
e-mail: thomas.bangemann@ifak.eu

M. Riedl
e-mail: matthias.riedl@ifak.eu

S. Karnouskos
SAP, Karlsruhe, Germany
e-mail: stamatis.karnouskos@sap.com

R. Camp
FluidHouse, Jyväskylä, Finland
e-mail: roberto.camp@fluidhouse.fi

O. Carlsson
Midroc Electro AB, Stockholm, Sweden
e-mail: oscar.carlsson@midroc.com

S. McLeod · R. Harrison
University of Warwick, Coventry, UK
e-mail: c.s.mcleod@warwick.ac.uk

R. Harrison
e-mail: robert.harrison@warwick.ac.uk

A. W. Colombo
Schneider Electric, Marktheidenfeld, Germany
e-mail: armando.colombo@schneider-electric.com

A. W. Colombo
University of Applied Sciences Emden/Leer, Emden, Germany
e-mail: awcolombo@technik-emden.de

P. Stluka
Honeywell, Prague, Czech Republic
e-mail: petr.stluka@honeywell.com

A. W. Colombo et al. (eds.), *Industrial Cloud-Based Cyber-Physical Systems,*
DOI: 10.1007/978-3-319-05624-1_2, © Springer International Publishing Switzerland 2014

from the latest state-of-the-art systems within the respective domain. While investigating the introduction of service-oriented architectures to automation, and even down to the shop floor, one has to consider latest standards, proofed technologies, industrial solutions and latest research works in the automation domain. This chapter tries, without any claim to completeness, to provide a short summary of today's situation and trends in automation.

2.1 Architecture of Production Systems

Several efforts to date have been directed towards defining structural and architectural aspects of production management systems. The most popular and applied in practice are the definitions set up within the ISA-95/IEC 62264 [21] standard. Typically, today's production systems (factory and process) are structured into a five-level hierarchical model (as depicted in Fig. 2.1). Besides this hierarchical, well-known model, IEC 62264 defines a manufacturing operations management model (like production control, production scheduling, maintenance management, quality assurance, etc.), which is not as popular, but implicitly represented by real installations.

The standard defines functions mainly associated to levels 3 and 4, objects exchanged and their characteristics and attributes, activities and functions related to the management of a plant, but does not specify about the implementations (tools) hosting these specific operations nor the precise assignment to one of the levels 2, 3 or 4. Realisations depend on individual customer needs and the tool manufacturer's strategies. For instance, maintenance management operation may typically be assigned to a Computerised Maintenance Management System (CMMS), a manufacturing execution system—both being typical Level 3 tools—but also to an Enterprise Resource Planning (ERP), dedicated to Level 4, or a Distributed Control System (DCS) that can be found at Level 2. Borders between these systems become floating.

Individual operations can be assigned to different specific manufacturing operations management areas—production operations management, quality operations management, maintenance operations management or inventory operations management. Having a look into these areas, individual activities (like resource management, detailed scheduling, dispatching, tracking, analysis, definition management, data collection, execution management [21]) can be identified to be executed within single or distributed sources. These functions can be implemented using different technologies. Currently, there is no standardisation regarding technologies to be used for implementing these functions.

2.2 Data Flow Within Automation Systems

The ways of communicating between the levels are different. Levels 1 and 2 are commonly connected through either point-to-point cabled solutions (4–20 mA current loop) or through fieldbuses (Modbus, Profibus, etc.). Ethernet and serial connections

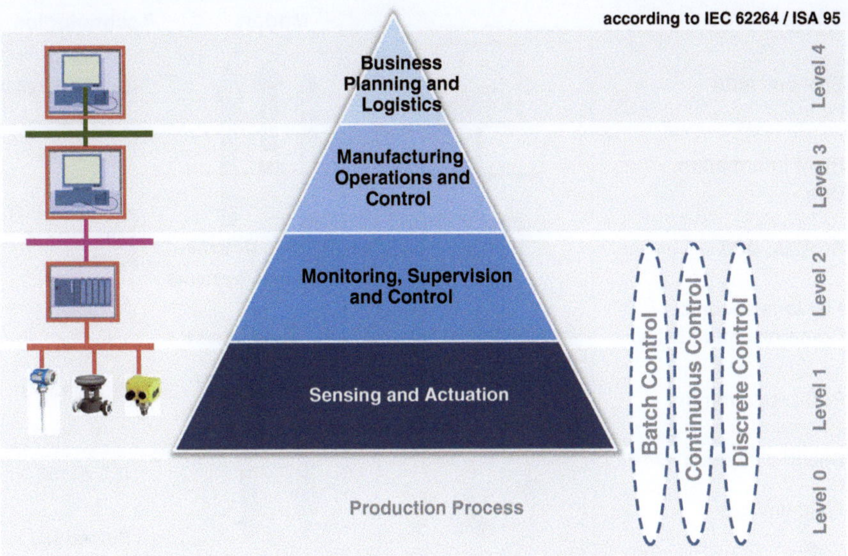

Fig. 2.1 Functional hierarchy according to (IEC 62264-3) [21, 39]

are used to an increasing extent as well. Fieldbuses and Ethernet can give an impression of a standard solution but the data exchange protocol on top of them is often proprietary, which leads to vendor lock-in. Some vendors start with a standard (electrical) interface but use a different non-standard connector, another kind of vendor lock-in. Because of this, end-users often must buy adapters, e.g. a converter to connect the serial port on the device to a port on the control system.

Figure 2.2 highlights some of the diversity of interfaces between the different levels and tools, which may even be distributed across the life cycle of a production system [25]. Profibus, Modbus or Foundation Fieldbus can give an impression of a standard solution. Fieldbuses standardise how to communicate; for instance, in order to configure a Profibus master to communicate with a slave, configuration files called GSD are required. These files specify the supported transmission speed and size of supported data buffers. GSD files can also hint about the interpretation of data. Additionally, semantics of data may be defined within device profiles, as done for Profibus PA or Foundation Fieldbus [13].

Monitoring of processes and automation equipment is an inherent pre-condition for keeping the production process alive and hopefully at near-optimal conditions to fulfil the business goals in the short, medium and long terms. It has to be guaranteed that data are provided:

- to the right application,
- in the quality (right semantics and syntax) needed for the consuming application,
- in right time (real-time) and sequence.

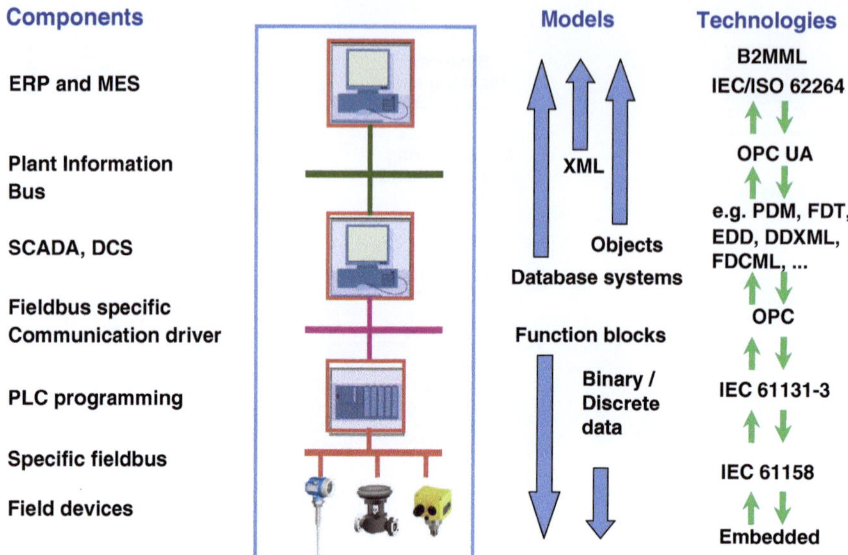

Fig. 2.2 Diversity of data and interfaces

Different applications raise specific requirements about the provision of data. Specifically for closed-loop control, data today must be retrieved in a cyclic manner. These sample times must be in a range that is suitable to the time-constraints of the controlled process. For that purpose, within a DCS, data are either polled internally from the DCS IO-cards, e.g. while accessing field devices through drilled lines supporting standardised signals (e.g. 4–20 mA analogue signal), or retrieved from remote-IO components via digital communication, following appropriate sample times, as described above.

Accessing process values within field devices through fieldbus communication is mainly done in a polling-based manner, e.g. Profibus with token-passing bus access, or based on the Publisher-Subscriber principle following configured cycle times (as done for Foundation Fieldbus). Transmitting data through digital protocols allows the association of status information (process and/or device related) to the process value. For instance, with Profibus PA communication, analogue process values are typically each transmitted as a Floating Point value associated with an 8-bit status in a single data structure each time a value is transmitted.

Considering the example of Profibus PA, the status Byte contains general information about device status, limit crossing of the process values measured, the validity of the process value as well as information to indicate maintenance demand. In cases when a failure is indicated, additional detailed information can be retrieved from the field device by individual a-cyclic requests. This construction of data allows interpretation by different types of applications:

- The process value, validity and limits are useful for the control application itself.
- This information will also be useful for supervision applications.
- Device status information is specifically needed for maintenance applications (Plant Asset Management).
- Production management applications will operate on more condensed data, representative of the production output. Such information typically is built by PLC or DCS based on information described above.

2.2.1 Use of Data for Supervision

SCADA deals with the gathering of data in real-time from remote locations in order to control and monitor the process, including data aggregation and presentation to the user. SCADA is commonly used in a broad range of application fields, like power plants as well as in oil and gas refining, telecommunications, transportation, and water and waste control, to mention a few. A typical SCADA system, as roughly depicted in Fig. 2.3, consists of several subsystems [23, 26] notably:

- A Human–Machine Interface (HMI) where the information is depicted and is used by human operators to monitor and control the SCADA linked processes.
- A computer which does the monitoring (gathering of data) as well as control (actuation) of the linked processes.
- Remote Terminal Units (RTUs) that are collecting data from the field (deployed sensors make the necessary adjustments and transmit the data to the monitoring and control system).
- Programmable Logic Controllers (PLCs) that are used as an alternative to RTUs since they have several advantages (like ability to deploy and run control logic) over the special-purpose RTUs.
- A communication infrastructure connecting all components.

SCADA systems include hardware and software components. The hardware gathers and feeds data into a computer that has a SCADA software installed. The software in a computer then processes these data and presents it in a timely manner. SCADA also records and logs all events into a file or sends it to a user terminal. These user terminals come in the form of Human–Machine Interface (HMI) or User Interface (UI) displays that allow the system to show data and warn when conditions become hazardous by generating alarms. Lastly, SCADA systems must ensure data integrity and appropriate update rates. Development of SCADA standards by industrial user groups and international standardisation bodies has allowed increased 'interoperability' of devices and components within SCADA systems [14]. Open protocols allow equipment from multiple vendors to communicate with the SCADA host. Many standards and specialised protocols exist with specific features.

Standards defining programming methods like IEC 61131-3 allow systems engineers to reuse code for logic operations and move easily between configuration interfaces. At the SCADA host level, the Open Connectivity via Open Standards

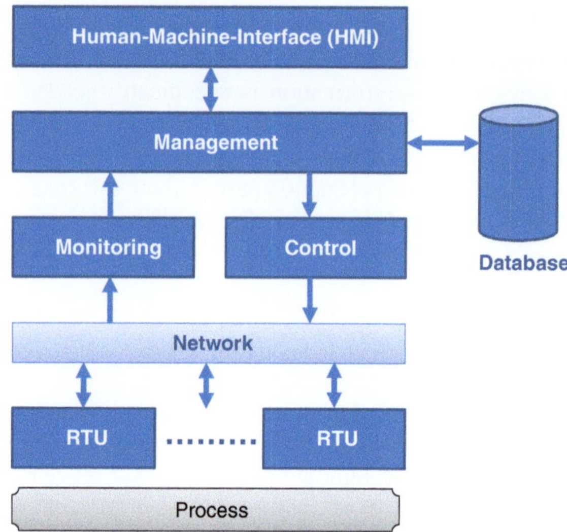

Fig. 2.3 Typical software architecture for a SCADA system

(OPC, previously OLE for Process Control) series of standards specifications have been widely accepted. Originally based on Microsoft's OLE Component Object Model (COM) and Distributed Component Object Model (DCOM) technologies, the specification defines a standard set of objects, interfaces and methods for use in process control and manufacturing automation applications to facilitate interoperability.

The OPC Foundation comprises a large group of vendor representatives dedicated to ensuring interoperability in industrial automation systems. The latest generation of SCADA system hosts the use of these OPC standards to provide advanced connectivity to user clients. The latest developments in OPC Foundation (www.opcfoundation. org/UA) denote: 'the new OPC Unified Architecture (OPC-UA) that is the next generation OPC standard (IEC 62541) that provides a cohesive, secure and reliable cross-platform framework for access to real time and historical data and events'.

These standards allow communications not only over serial links for dedicated communication channels, but also transfer of SCADA data over Ethernet with a TCP/IP protocol stack for Wide Area Networks (WANs) or Local Area Networks (LAN). Therefore, it is understood to benefit from an advanced high-speed, peer-to-peer communication service as well as improved device interoperability for process monitoring and automation, without the need for high cost of integration.

2.2.2 Use of Data Within Process Control Architectures

After decades of analogue single-loop controls, the early minicomputers started the transition to digital control systems in the 1960s. The Distributed Control System

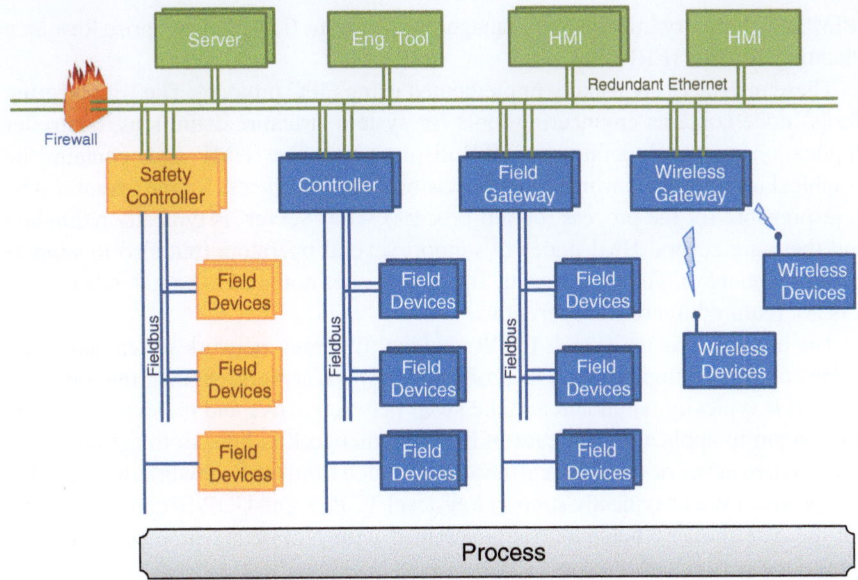

Fig. 2.4 State-of-the-art distributed control system

(DCS) was introduced at roughly the same time (1975) by Honeywell (TDC 2000) and Yokogawa (CENTUM). This was partly due to increased availability of microprocessors. The early DCSs were designed using proprietary hardware and software. The latest DCSs contain lots of Commercial off-the Shelf (COTS) components and IT standards are utilised whenever possible.

Today's state-of-the-art DCS has several nodes for different purposes as depicted in Fig. 2.4. The nodes are able to communicate using high-speed networks. Some of the nodes and networks are redundant and can tolerate single failure. The level of redundancy depends on industrial requirements, e.g. in the food and beverage industry the level of redundancy is quite limited while in the petrochemical industry almost all components are redundant. The DCS architecture is able to support a free combination of redundant and non-redundant components. It is also a very scalable architecture supporting all kinds of systems from very small (PC and some I/O channels) to very large and distributed systems (consisting of tens of thousands of I/O points and thousands of control loops). One of the goals in these systems is to secure the deterministic behaviour of the system at all levels in all circumstances.

The highest level nodes are 'Server', 'Engineering Tools' and human–machine interface 'HMI'. Today these are almost always PCs with Microsoft Windows operating system. The 'Server' contains all the configurations that are needed in the other nodes at runtime or in cold-start situations. It typically also contains data history collections, master alarm lists and perhaps interfaces to some other systems. These systems can be other DCS systems, Programmable Logic Controllers (PLC), Manufacturing Execution Systems (MES), Process Information Management Systems

(PIMS), Laboratory Information Management Systems (LIMS), Enterprise Resource Planning systems (ERP), etc.

These interfaces are usually implemented using OPC protocol. The 'Engineering Tools' node contains engineering tools for system structure definitions, controller applications, network definitions, HMI displays, etc. The 'HMI' node contains the graphical user interface which provides visibility to the process for the operator who is responsible for the process (or sub-process). The 'Server' is typically redundant and there are several HMI nodes to support several operators (but also to support HMI redundancy). The Engineering Tools node does not need to be redundant since it is not required in normal operations.

The highest level network is the 'Redundant Ethernet' network which takes care of the communication between controllers, gateways, servers, engineering tools and HMI. It is typically redundant and the swap between active and passive network is transparent to applications in case of hardware/network failure. Both networks use independent network switches and these are isolated from other networks by firewalls. The protocol stacks typically support low-level TCP/IP and UDP/IP communication but the deterministic behaviour is guaranteed with proprietary protocols that take care of the network utilisation.

The 'Controller' node is an important node in the system. It is where the most important control algorithms (closed and open loop) and logic are running. These nodes use proprietary hardware and software environments. The hardware supports some kind of non-volatile memory and high-speed redundancy. In many cases it is also designed to survive in harsh environments. The execution environment runs on a hard real-time operating system executing typically function block configuration but also other programming languages (in a time-constrained manner).

The controller is either connected directly or through a 'Field Gateway' to the fieldbuses. The fieldbuses are based on (mostly de facto) standards. The most popular fieldbuses are Foundation Fieldbus (H1 and HSE), PROFIBUS (DP and PA) and Ethernet-based PROFINET. The fieldbuses and field devices can be redundant or non-redundant. The protocols used in these fieldbuses can guarantee the deterministic behaviour when delivering critical data. The less time-critical data (e.g. diagnostics data) is transferred in the remaining time slots. It is also possible to add digital communication to field devices that are connected using traditional analogue 4–20 mA cables using the HART protocol. It is also possible to integrate wireless devices into the DCS architecture using (redundant) 'Wireless Gateway'. With these devices it is more difficult to guarantee the deterministic behaviour because of the less robust media. Several protocols are available, including WirelessHART, which maintains compatibility with existing HART devices, commands and tools.

In some industries, special industrial safety systems are required to protect humans, plants and the environment in case the process goes beyond the control limits. These are also part of the DCS architecture. The 'Safety Controller' contains special redundant hardware which is Safety Integrity Level (SIL) certified.

The controllers are able to transfer data to each other (peer-to-peer communication). These data are typically transferred cyclically with defined time intervals but can be also event based. The communication protocols at controller level guarantee the deterministic behaviour and in many cases data subscriptions are used. The alarms are always event based. The controllers (and other nodes in the system) generate alarms for the operator and these typically require human acknowledgement. The alarm list is maintained by the server and shown on the HMI nodes. The data for the HMI displays (graphical view of the process) show the live data that is transferred from the controllers. Usually, the data are only transferred to displays that are currently switched on.

The software architecture inside the distributed control system is still based on object-oriented principles. Services are available but in many cases they are not created as granular components. Also, the interfaces are typically used for direct (local) method calls or direct data access rather than standards-based open remote interfaces. Online service discovery is also limited. Moving to SOA in distributed control systems would clearly bring architectural benefits and ultimately benefits for the users through services being more open, easy to find and accessible for external applications. It would also simplify the development and maintenance of the distributed control system and support new capabilities.

2.2.3 Use of Data for Production Management

Enterprises are moving towards service-oriented infrastructures that bring us one step closer to the vision of 'real-time enterprises' [27]. Applications and business processes are modelled on top of and using an institution-wide or even cross-institutional service landscape. For any solution to be easily integrated in this environment, it must feature a service-based approach.

One can realise a 'real-time enterprise' via strong coupling of the enterprise concepts domain and the device-level service domain. Nowadays, there is multi-step cooperation between the two layers, which in practice translates into the coupling of Enterprise Resource Planning (ERP) with Manufacturing Execution System (MES) and Sistributed Control System (DCS). By integrating device-level services with higher level enterprise services, timely information can flow to business processes and enhance existing applications.

As the whole enterprise is seen as a complex ecosystem, every process may affect several others in the system and therefore need to be managed in an integrated way. This includes:

- Warehouse and production management—Management of inventory across multiple warehouses, tracking of stock movements and management of production orders based on material requirements planning.
- Customer relationship management.
- Purchasing—Automation of procurement process from purchase order to vendor invoice payment.
- Reporting—Real-time information with detailed reports.

There are several IT systems that exist in the factory or plant floor today and data that are collected at various levels. At the lowest level is SCADA systems as repositories of field real-time massive data as they collect data from the PLCs and sensors that are connected to the machinery on the factory or plant floor. At the next higher level are MES that track all customer orders, schedules, labour, resources and inventory across the production line by shift. At the uppermost ERP and other enterprise solutions like Supply Chain Management (SCM), etc., plan and record transaction data to measure variance against set performance targets, etc.

Unfortunately, in many manufacturing companies today, these three layers are still not fully integrated. As a consequence, companies often employ large numbers of people to punch in or import redundant production batch data from their MES to their ERP systems. This is not only a wasteful and costly exercise but also introduces human errors in the data entry process. Even if done in an automatic way, this usually includes huge delays (sometimes in days), which prohibits the managers from getting a real-time/right-time picture of factory performance, variance from set targets as well as order/materials/machine/labour/quality/maintenance exceptions and issues that may arise in the factory. The latter may be translated into lost opportunities, e.g. failure to optimise production or even unhappy customers due to delayed shipments.

While the SCADA and MES layers tend to be integrated at most companies, it is equally likely that the heterogeneity of this environment comprising home-grown, legacy and point applications from multiple vendors with differing architecture platforms may result in disconnections in this layer as well. This tends to further exacerbate the problem.

The business implication of any exception or the ability to compare actual manufacturing performance against set targets is not evident until MES data and exceptions from the factory floor hit the ERP system. ERP in essence, if integrated seamlessly with the factory MES layer, provides the business context for manufacturing transactions, exceptions and issues captured on the factory floor. The bottom-line implication for manufacturers is that the disconnect between the Shop Floor (Factory MES) and the Enterprise Top Floor (ERP) costs them millions of Euro through waste, reject, re-orders, expedites, preventable material/machine/labour/quality issues that are detected too late, for enterprises to proactively resolve them.

Based on these considerations one can identify distinct directions towards the organisational structure of a production site and the topological or architectural characteristics. From the organisational point of view, the business is typically structured in a similar way to the levels and operations defined by IEC 62264, however, it might be better to express this in the opposite manner, i.e. that the standard is following what has been developed over the past years. Structures, skills, responsibilities, professions, education, etc., have been established focussing these organisational matters. It is questionable if, and how fast this may change in the future.

2.3 Integration Technologies Between Layers and Applications

Today, integration of Legacy Systems into new state-of-the-art systems has becoming an elementary task for each solution provider or engineering company. Legacy systems undergo continuous changes and modifications due to even more frequently changing requirements imposed by market needs. Normally, this progressively causes a significant increase in the complexity of existing systems [7]. The main problem with the integration process is the heterogeneity among systems. The heterogeneity issue [22] can be divided into:

- Technological Heterogeneity, e.g. different hardware, operating systems, communication protocols for accessing data and programming languages.
- Semantic Heterogeneity (e.g. the same names of data sources but different meaning or different names associated with the same meaning).

From the software architecture point of view, in order to integrate legacy systems, the role of each subsystem or component that is to be integrated has to be defined along with the interfaces and building object wrappers for each subsystem. An integration approach, where the system developer is required for knowing the internals of the legacy system is known as White-Box approach and an integration approach that only requires knowledge of the external interfaces of the legacy system is known as Black-Box approach [7, 11]. In order to integrate legacy devices into state-of-the-art automation systems, legacy adapters can be used, being composed of [31]:

- State-of-the-Art Interface Layer (required to communicate with the state-of-the-art system, configuration capabilities have to be provided),
- Integration Layer (used for protocol transformation, data and semantics transformation; configuration capabilities have to be provided),
- Legacy Systems Interface Layer (provides the communication capabilities for exchanging data with legacy components, configuration capabilities have to be provided).

There are different ways to integrate legacy systems using adapters, e.g. by utilising gateways or mediators. Besides these general concepts, specific technologies and concepts for integration of data are used or approached in today's automation systems, e.g. Electronic Device Description (EDD), Field Device Tool (FDT), Field Device Integration (FDI), OPC Unified Architecture (OPC-UA).

2.3.1 Integration Using Gateways and Mediators

Using gateways is a well-proven concept for integrating/connecting devices, attached to different networks. It is used to transform protocols as well as the syntax of data. Semantic integration is harder to achieve. Nevertheless, it is possible to do transformation between data centric approaches, as typically followed by fieldbus concepts and service-oriented approaches.

A gateway, as defined in the FP6 SOCRADES [25] and FP7 IMC-AESOP projects, is understood to be a device that controls a set of lower level non-service-enabled devices, each of which is exposed by the gateway as a service-enabled device (as depicted in Fig. 2.5). This approach allows the gradual replacement of limited-resource devices or legacy devices by natively service-enabled devices without

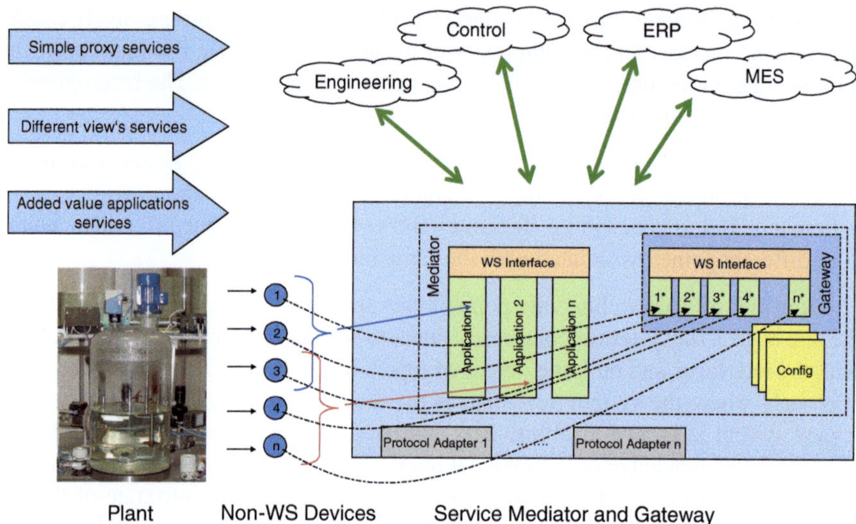

Fig. 2.5 Gateway and mediator concepts for integration of devices [1]

impacting the applications using these devices. This approach is used when each of the controlled devices needs to be known and addressed individually by higher level services or applications.

The mediator concept is based on the elaboration of the gateway concept, while adding additional functionality to the gateway. Originally meant to aggregate various data sources (e.g. databases, log files, etc.), mediator components have evolved with the advent of Enterprise Service Bus (ESB) [17]. Service mediators are now used to aggregate various services in SOA. As such, a mediator can be seen as a gateway, except that it can hide (or surrogates) many devices, not just one. However, service mediators also go beyond gateways since they introduce semantics in the composition. Mediators aggregate, manage and eventually represent services based on some semantics, e.g. using ontologies.

2.3.2 Electronic Device Description

An Electronic Device Description (EDD) is based [20] on a formal language called Electronic Device Description Language (EDDL). This language is used to describe completely and unambiguously, what a field instrument looks like when it is seen through the 'window' of its digital communication link. EDD includes descriptions of accessible variables, the instrument's communication related command set and operating procedures such as calibration. It also includes a description of a GUI structure which a host application can use for a human operator. The EDD, written in a readable text format, consists of a list of items ('objects') with a description of the features ('attributes' or 'properties') of each.

The major benefit of EDD for device suppliers is that it decouples the development of host applications and field devices. Each designer can complete product development with the assurance that the new product will interoperate correctly with current and older devices, as well as with future devices not yet invented. In addition, a simulation program can be used to test the user interface of the EDD, allowing iterative evaluation and improvement, even before the device is built.

For the user, the major benefit is the ability to mix products from different suppliers, with the confidence that each can be used to its full capacity. Easy field upgrades allow host devices to accept new field devices. Innovation in new field devices is encouraged. The EDD is restricted to the description of a single device and use in a mostly stand-alone tool, preferably for commissioning the field devices. Due to the nature of EDD such tools are based on interpreter components suitable to the EDDL.

Software tools for automation are complex, and implement a lot of know-how. The number of sold products is relatively low in comparison with office applications. The definition of standardised device description languages increases the potential users of such tools and also encourages the use of fieldbus-based automation.

2.3.3 Field Device Tool

In order to maintain the continuity and operational reliability of process control technology, it is necessary to fully integrate field devices as a subcomponent of process automation [36]. To resolve the situation, the German Electrical and Electronic Manufacturers' Association (ZVEI) initiated a working group in 1998 to define a vendor-independent Field Device Tool (FDT) architecture, the specification of which is maintained and refined inside the FTD Group (www.fdtgroup.org).

This FDT concept defines interfaces between device-specific software components (DTM—Device Type Manager) supplied by device manufacturers, and engineering systems supplied by control system manufacturers. The device manufacturers are responsible for the functionality and quality of the DTMs, which are integrated into engineering systems via the FDT interface. With DTMs integrated into engineering systems, a unified way of creating the connection between engineering systems (e.g., for PLC applications) and currently inconsistent field devices becomes available. The FDT specification defines what the interfaces are. DTMs act as bridges between the frame-application and field devices. Several technical documents on FDT summarise the available features (more info available at www.fdtgroup.org/technical-documents).

2.3.4 Field Device Integration

Looking into the market situation, it can be noticed that both aforementioned technologies for device integration, i.e. EDD and FDT, are competing on the market [16]. On one hand, benefits of EDDL such as robustness, independence from the operating system and backward compatibility are promising characteristics for the system

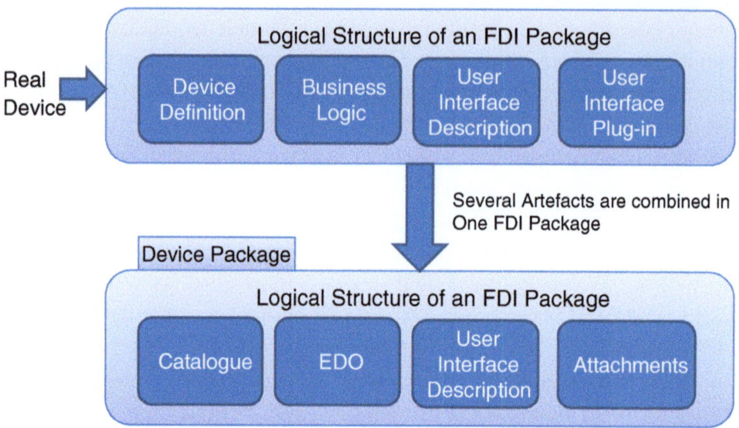

Fig. 2.6 Structure of an FDI device package [16]

integrator or the end-user. On the other hand, the FDT approach provides potential to allow the device vendors to represent their brand label, realising highly sophisticated user interfaces to the end-user. FDT components may be easily plugged into a DCS or other commissioning and operations management tools, which is seen by the user as a useful service.

The system providers have to handle more and more complex systems. Such systems will be less homogeneous and more distributed, having different network technologies, including gateways between them or requiring worldwide online access. Although existing solutions may offer such features they will often be proprietary. EDDL and FDT are the basis of Field Device Integration (FDI) [16], which is targeting to provide a way of migration of both technologies (EDDL and FDT). It is intended to take advantage from the more promising concepts of both technologies.

In FDI the device is represented by an FDI device package, Fig. 2.6, and covers all information needed for the integration of the field device into the automation system. The device vendor provides the FDI device package. It replaces the EDD or DTM and consists of several components as shown in Fig. 2.6, but the end-user now has to install only one file—the FDI device package—in the system. Thus, this is a significant improvement in handling such a complex information pool.

The FDI device package consists of logical blocks such as device definition, business logic, user interface description and user interface plug-in [16]. Device definition describes the parameters of the device and its internal structure, e.g. blocks or modules. Business Logic ensures the consistency of the device parameters (this means also the consistency of the device model, see above). Examples of such consistency rules are dynamic conditions or relations between parameters. Thus, parameter values could be changed depending on the device status/device configuration. GUI elements could be available as descriptive elements (user interface descriptions) or as programmed components (user interface plug-ins).

2.3.5 OPC: Unified Architecture

Classical OPC is a technology widely used as a basic communication platform for integrating data for supervision and control purpose based on information models defined. Many products (such as PLC, DCS and SCADA devices) exist on the market supporting OPC server or client components. During the last years the original OPC specifications, based on Microsoft COM/DCOM, were replaced by new interoperability standards, such as Web services. Consequently, the OPC Foundation published the OPC Unified Architecture (OPC-UA) [32].

The transition towards this unified architecture started with the development of the OPC XML DA specification, which introduces the use of XML, thus allowing the flow of information beyond corporate firewalls and permitting cross-platform connectivity via Simple Object Access Protocol (SOAP) and Web services through the Internet [19]. The limitations of OPC-UA however, are mainly evident at the factory level, namely at the device level. While OPC-UA allows the integration of process control devices with SCADA and even MES systems, the information offered by low-level devices can only be accessed through process control systems. In order to further expand the reach and flow of information, device integration standards such as Field Device Tool (FDT) and Electronic Device Description Language (EDDL) can be used [19].

Several technology supporting organisations—such as PROFIBUS International (PI), Fieldbus Foundation (FF), HART Communication Foundation (HCF), or others—started investigating the potential use of OPC-UA to take advantage of this basic technology. As an example, PLCopen and OPC Foundation are undertaking common activities to jointly define a common information model. Information models have been developed for Electronic Device Description (EDD) and now also for IEC 61131 PLC. This development ensures that field devices that are described in EDD and in future that are represented by PLC proxies can be accessed by OPC-UA Web services (more info is available at www.plcopen.org).

OPC-UA uses client-server architecture with clearly assigned roles. Servers are applications that expose information following the OPC-UA information model, where each server defines an address space containing nodes of the OPC-UA model. These nodes represent real physical or software objects. Clients are applications retrieving information from servers by browsing and querying the information model. Both types of applications can be developed using an API that isolates the application from the communication stack. Figure 2.7 gives an overview of the flexibility and extensibility of the OPC-UA architecture.

Interoperability and adaptability of the standard are reachable through several complementary features of OPC-UA:

- Extensible object model;
- Rich set of services;
- Scalability;
- Reliability, Redundancy and Performance;
- Security;

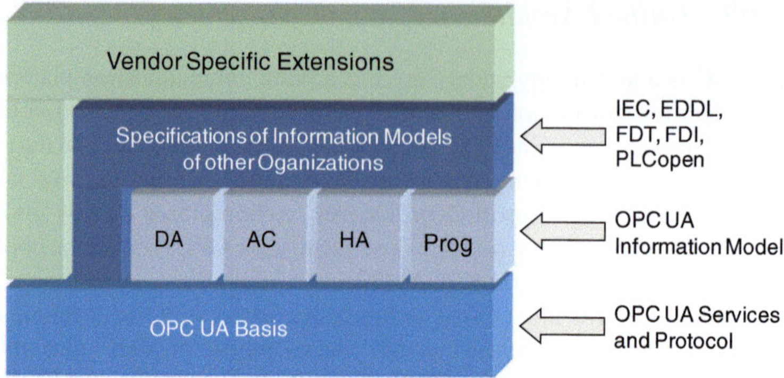

Fig. 2.7 Overall OPC-UA architecture [32]

- Backwards compatibility;
- Standardisation at the protocol level;
- Isolation of the application from the communication stack through the client or server API.

2.4 Engineering of Production Systems

There is an on-going trend towards higher levels of automation in process control systems [15, 31] with increasing levels of autonomy in control and monitoring. Today's automation/business systems are moving to a 'Smart' environment such as smart devices, smart systems, smart organisations and smart cities, where Smart may be defined as systems that exhibit (i) extended functionality, (ii) multi- functionality, (iii) self-diagnosis, (iv) configurability and (v) connectivity.

With the increased use of Commercial off-the Shelf (COTS) technologies, the network infrastructure of the DCS and network architecture for plant information become increasingly interdependent. The prevalence of Ethernet at every level of an organisation, especially in green-field sites provides shop-floor systems with the infrastructure for data acquisition, analysis and integration with other enterprise systems [37]. This also creates problems with the proliferation of data, which requires integration and management.

Tools and methods are required to manage this and make complex time-dependent data integrated from disparate sources available to other systems within the enterprise in a consistent manner. Users of these systems are becoming more demanding too, it is expected that timely data should be available 'anywhere, anytime and on multiple platforms' (e.g. mobile and web devices). Additionally, users will expect systems to be richer not only in content but graphically too, and they will expect more interactive graphical systems with emphasis placed on design of the user interface as well as the functionality being offered [29].

The current trend in manufacturing system design tool development consists in merging system mechanical and control design software in a single environment in order to break the communication barrier that commonly exists between mechanical and control engineers and which translates into difficulties to coordinate two complex, but separated design processes. This approach is dominated by Siemens (Process Simulate) and Dassault Systems (Delmia) providing solutions that can potentially take CAD models and provide 3D kinematic simulations to validate the mechanical design and engineering process, and generate code for deployment on PLCs [18]. A related approach that has gained popularity uses Winmod for modelling control behaviour and Invision for 3D modelling, which allows the virtual commissioning and simulation of automation systems [33].

Traditional shop-floor applications are likely to be superseded by cloud-based applications (where hardware control is not an issue), and with the introduction of software-as-a-service ('SaaS') models, it means that software will be less hardware dependent and more dynamic in nature as service upgrades should happen without shop-floor intervention [6, 28].

Smart network attached devices are becoming more and more powerful and cheap to produce; the expected resultant explosion in these devices will lead to more widespread use of DCS, where devices will cooperate in a peer–peer way to meet the system goals [8]. These devices will drive engineering tools and methods to handle the building and development of systems as a set of cooperating modules or components whose application logic is either centralised and the device behaviour is orchestrated, or the application logic is distributed to the devices and the overall behaviour is choreographed. In either method, tools are capable of integrating devices from different vendors and domains (e.g. business, external and automation components). One promising methodology for achieving this is the use of AutomationML (www.automationML.org), which is described as a neutral data format for automation engineering.

In addition to these design and development tools, engineering tools are required to support the complete life cycle of an automation system. In many cases, these virtual engineering approaches are used to create automation systems and provide visualisations that can be used as a catalyst for communication and understanding between disciplines (such as mechanical, control and safety engineering) and even the supply chain, but once the system is commissioned these models are not kept up-to-date with changes that occur during its life, due to the time and cost associated with maintaining the original models.

There is a requirement for lightweight visualisations that may be used to aid in diagnostics and maintenance; these tools should be directly linked to the automation systems such that changes may be quickly and simply made in the engineering tools validated and then deployed directly on the system, or when changes are made directly on the physical system when the model will reflect these changes implicitly. In this way greater return on investment in modelling and simulation can be achieved.

If this trend of building heterogeneous systems continues, systems will become more modular and componentised. This should enable systems to be built from a blend of the best custom-built apps and off-the shelf-components, which could make the market more open and more competitive. The introduction of these will be dependent on the ability of such systems to be maintained effectively and to ensure that the production downtime is still kept to a minimum. Acceptance of such technologies is likely to depend on familiarity of control representation (e.g. ladder, timing/Gantt chart, function block diagrams), such that engineers will be able to understand and maintain them using their core knowledge.

Advances in active tagging result in direct or indirect tagging of devices, work pieces, employees, etc., and as they become cheaper and more widely used, future automation systems should be capable of using this information and integrating it with control to enhance performance (e.g. live inventory control), safety (e.g. employee tracking) and maintenance (e.g. location of mechatronic devices). In conclusion, technological and infrastructural advances in automation system design manufacture and deployment is happening rapidly, however, engineering tools capable of effectively supporting and exploit these advances are severely lacking or fragmented. The challenge is therefore to provide engineering tools and effective interoperability between such tools for the next generation of DCSs.

2.5 Towards SOA-Based Automation

Among the biggest challenges faced by manufacturing enterprises are the constant demands to change their processes and products and still be able to manage the inherent complexity in all levels of their production environment. In order to provide the IT support needed to cope with these challenges, appropriate ways of designing automation software systems are required. As a consequence, factory automation providers are integrating the SOA approach in their solutions for Manufacturing Execution Systems (MES), Enterprise Resource Planning (ERP) or Enterprise Asset Management (EAM) systems.

However, many challenges remain when applying the service technology to the shop floor devices characterised with limited resources and real-time requirements. At this level, the interactions are still carried out using different fieldbus and industrial Ethernet protocols with restricted interoperability across technology borders. This limits the ability to enforce plant-wide, seamless integration of processes and services leading to complex systems for monitoring and control that are heavily dependent on the interactions with various resource constrained shop floor devices such as sensors and actuators.

2.5.1 Building Service-Based Infrastructures

To overcome this situation and to address integration of very large numbers of subsystems and devices (including field level devices) within a harmonised networking architecture, several European collaborative projects such as IMC-AESOP [26], SIRENA [2], SODA [12], SOCRADES [9, 38], etc., investigated Web services at the device level and integrated these devices with MES and ERP systems at upper levels of an enterprise architecture [10, 24, 27]. The first results shown in pilot applications running in the car manufacturing, electromechanical assembly and continuous process scenarios have been successful [4], confirming that the use of cross-layer service-oriented architectures in the industrial automation domain is a promising approach. Additional examples, coming from the IMC-AESOP project are presented within Chaps. 7–10, highlighting the use of Web service technologies within the domain of control and monitoring of batch and continuous processes.

The FP6 SOCRADES project evaluated several SOA solutions, applicable at the device level in the context of manufacturing automation. The SOCRADES (DPWS based) solution was provided as a complete open-source software component, which was embedded in several devices and tools, and was demonstrated in electronic assembly demonstrators, continuous process control and in interoperability trials. A potential merger between DPWS and OPC-UA was also identified [3, 35]. Potential solutions were identified to reduce the costs of embedding DPWS in very simple devices. Generic and application Web services were identified, specified and implemented in prototype applications.

To overcome the often-poor integration between engineering methods and tools, IMC-AESOP looked at tools and methods established, or emerging, in the process control sector, plus applicable approaches from other domains relevant to an SOA-based engineering approach. The engineering requirements of large-scale process control systems were considered likely to be somewhat different from the smaller scale systems previously considered in SOCRADES, i.e. in terms of control and monitoring, traceability and integration with management systems, data acquisition and reporting, and system reliability and security [30].

The IMC-AESOP project considered the state of the art in engineering tool life cycle engineering capabilities and related user application requirements from the perspectives of:

- Monitoring;
- Control;
- Enterprise and management integration systems, e.g. application of SCADA and MES;
- SOA engineering methods, tool and the application of Web services;
- System visualisation, e.g. 2/3D system visualisation;
- Simulation methods, e.g. optimisation and key performance controls, prediction of system behaviours;
- Quality control;
- Environmental factors, e.g. energy optimisation.

Based on the findings it is considered that, in an SOA context, engineering applications of the future will need to:

- *Provide integration.* People and computers need to be integrated to work collectively at various stages of the product development and even the whole product life cycle, with rapid access to required knowledge and information. Heterogeneous sources of information must be integrated to support these needs and to enhance the decision capabilities of the system. Bi-directional communication environments are required to allow effective, quick communication between human and computers to facilitate their interaction.

- *Be heterogeneous*. To accommodate multi-vendor and multi-purpose software and hardware in both manufacturing and information environments.
- *Be interoperable*. Heterogeneous information and control environments may use different programming languages, represent data with different representation languages and models and operate in different computing platforms. Yet these subsystems and components should interoperate in an efficient manner.
- *Be open and dynamic*. It must be possible to dynamically integrate new subsystems (software, hardware or manufacturing devices) into or remove existing subsystems from the system without stopping and reinitialising the working environment.
- *Be agile*. Considerable attention must be given to reducing product cycle time to be able to respond faster to customer desires. Agile manufacturing is the ability to adapt quickly in a manufacturing environment of continuous and unanticipated change and thus is an essential component in manufacturing strategies for global competition. To achieve agility, manufacturing facilities must be able to rapidly reconfigure and interact with heterogeneous systems and partners.

The advantage of Service-Oriented Architectures (SOA) in the industrial automation domain are manifold including: device virtualisation using Web services; automatic composition, orchestration and configuration of distributed automation functions and systems by means of service-based applications; use of technologies at the research edge providing real-time and large-scale industrial automation and control applications. However, as identified by the SOCRADES project the significant benefits assume that several challenges will also be adequately addressed [38].

2.5.2 Virtualisation of Smart Embedded Automation Devices with Web services

Typical production equipment like transport units, robots, but also sensors, valves, etc., are considered as modules integrating mechanic, electronic, communication and information processing capabilities. This means that the functionalities of the modules are exposed via Web services into a network, as depicted in Fig. 1.2 [5]. Embedding Web service protocols into the automation device, e.g. DPWS or OPC-UA [34] allows the transformation of traditional industrial equipment into the nodes of an information-communication-network. Such nodes will be able to expose and also to consume 'Services'. Moreover, depending on the position and inter-relation of such nodes to other nodes of the network, it becomes necessary to compose, orchestrate and/or choreograph services.

The virtualisation of a mechatronic module transforms it into a unit able to 'collaborate' with other units. That is, a module that communicates with others, exposing or consuming 'Services' related to automation and control functions. Recent trends

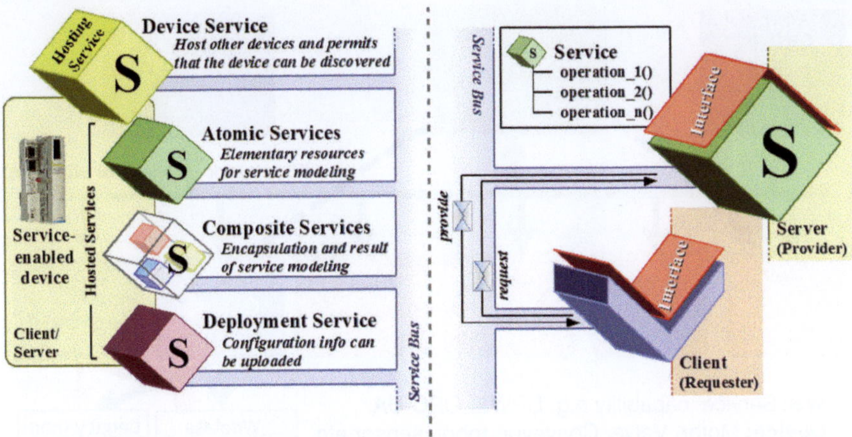

Fig. 2.8 Web service classification for SOA-compliant smart embedded device

in the technology developments associated to automation devices facilitate the virtualisation: Web service protocols are now embedded into a chip, integrated into industrial automation and control devices.

Different specifications of a collaborative mechatronics module and the corresponding smart automation device are virtualised and the resultant 'Services' can initially be classified according to the position and offered functionality of the smart device. Figure 2.8 shows an initial classification of the 'Web services' that will be exposed to the network and will immediately be ready to be consumed/requested from other nodes of the SOA-based network.

2.5.3 Configuring a Shop Floor as an SOA-Based Collaborative Automation Network

A shop floor composed of smart embedded devices that follow the specifications already discussed appears as a flat automation architecture, where each component has a Web service interface and may take part in various orchestrations collaborating with other service-enabled devices and systems.

Within Fig. 2.9 the block with the denomination 'Service Orchestration' represents a module that is able to compose and orchestrate 'Services'. This logic function will be implemented in a centralised or distributed manner, depending on the kind of virtualised system. This means, orchestration (or even choreography) engines will be deployed into one or more smart automation devices, i.e., another SW component and processing engine inside the smart device. Devices are 'motors', 'valves', 'conveyors', 'storages', 'HMI', 'drives' and generally any mechatronic components with CPU-capability and embedded Web service stack. PLC and robot controllers can also be transformed into 'service producer/consumer' integrating Web service capabilities.

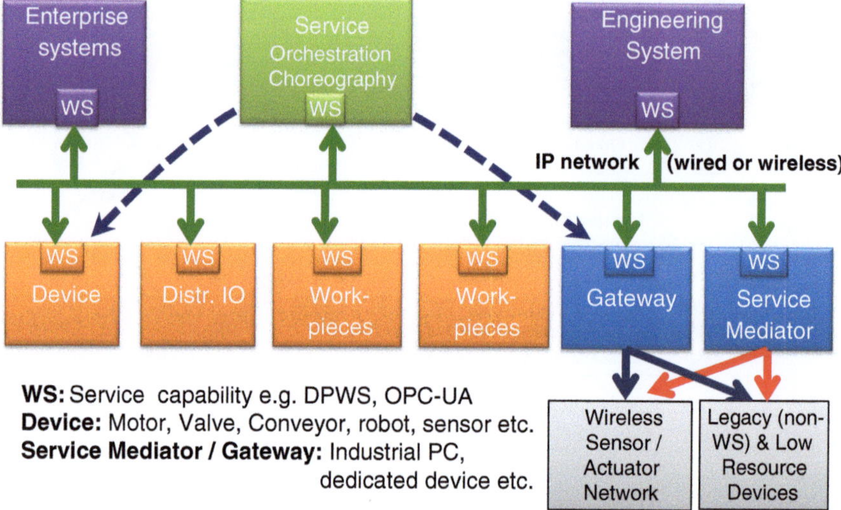

Fig. 2.9 Flat SOA-based technical architecture of production systems

One of the major outcomes of the Web service-based virtualisation of a shop floor is the possibility to manage the whole system behaviour by the interaction of Web services, i.e. exposition, consumption, orchestration, choreography, composition of the different kind of services exposed by the different SOA-compliant smart devices and systems.

A deeper analysis of the SOA-based automation systems shows that the SOA-based virtualisation, applied to an enterprise, makes a clear transformation (from the architectural point of view) of the traditional hierarchical ISA-95 compliant enterprise architecture into a 'logical' flat architecture [28]. This major and fundamental outcome of the Web service-based virtualisation of a shop floor relies on the fact that the 'Services', when they are exposed using the same Web service-based protocol, are directly consumed, composed and/or orchestrated in an independent way from the source (where these services are physically originated). A Web service exposed by the MES component (located in the ISA-95 Level 3) can immediately be composed with a Web service generated by a valve (located in the ISA-95 Level 1).

Topological and architectural characteristics are driven by user or application needs with respect to latest, proven or acceptable technological capabilities. IMC-AESOP proposes and follows the idea of establishing a service cloud fulfilling today's requirements for production management systems. The composition of the cloud is targeted towards the suitability of supporting IEC 62264 operations and activities. Thus, one may still keep the organisational aspects established in today's production systems, while migrating to a future SOA-based underlying architecture, exploiting the desirable capabilities inherent to SOA.

2.6 Conclusion

PLC, SCADA and DCS systems are the basis for monitoring and controlling industrial applications at lower levels within the plant hierarchy. Upper levels are dominated by MES and ERP systems. Information exchange at lower levels is characterised by a data-centric approach utilising industrial serial fieldbus systems or Ethernet-based communication supported by appropriate engineering concepts and tools. Diverse standardisation activities towards interoperability have been undertaken in the past, focussing individual device classes, programming concepts or communication capabilities of neighbouring levels. All these, as roughly introduced within this section, are widespread across industrial sectors.

The more complex, large and diverse applications become, limits are reached by existing technologies requesting improvements or even new technologies to be introduced. On the other hand, innovations may only be as large and introduced as fast, as the user is able and willing to adopt them. Consequently, every work towards challenging targets must start from the base-ground. This chapter was dedicated to give a brief, not raising any claim for completeness, overview of the state of the art in industrial automation as well as some progress actually monitored. Based on this, the following chapters will introduce the innovative results of the IMC-AESOP project.

Acknowledgments The authors thank the European Commission for their support, and the partners of the EU FP7 project IMC-AESOP (www.imc-aesop.eu) for fruitful discussions.

References

1. Bangemann Th, Suchold N, Colombo AW, Karnouskos S (2012) Die Integration Service orientierter Architekturen in der Automation. In: Automation 2012, Baden-Baden, 13–14 June 2012. VDI-Berichte 2171, S 333–336, VDI Verlag GmbH, Düsseldorf. ISBN 978-3-18-092171-6
2. Bohn H, Bobek A, Golatowski F (2006) Sirena—service infrastructure for real-time embedded networked devices: a service oriented framework for different domains. In: International conference on networking, international conference on systems and international conference on mobile communications and learning technologies. ICN/ICONS/MCL 2006, p 43. doi:10. 1109/ICNICONSMCL.2006.196
3. Bony B, Harnischfeger M, Jammes F (2011) Convergence of OPC UA and DPWS with a cross-domain data model. In: 2011 9th IEEE international conference on industrial informatics (INDIN), pp 187–192. doi:10.1109/INDIN.2011.6034860
4. Boyd A, Noller D, Peters P, Salkeld D, Thomasma T, Gifford C, Pike S, Smith A (2008) SOA in manufacturing—guidebook. Technical report, IBM Corporation, MESA International and Capgemini. ftp://public.dhe.ibm.com/software/plm/pdif/MESA_SOAinManufacturingGuidebook.pdf
5. Candido G, Jammes F, Barata J, Colombo A (2009) Generic management services for DPWS-enabled devices. In: 35th annual conference of IEEE on industrial electronics, 2009 (IECON '09), pp 3931–3936. doi:10.1109/IECON.2009.5415339
6. Chowanetz M, Pfarr F, Winkelmann A (2013) A model of critical success factors for software-as-a-service adoption. In: 7th IFAC conference on manufacturing modelling, management, and control, St. Petersburg, Russia

7. Chowdhury MW, Iqbal MZ (2004) Integration of legacy systems in software architecture. In: Specification and verification of component-based systems (SAVCBS) workshop at ACM SIGSOFT 2004/FSE-12
8. Colombo A, Mendes J, Leitao P, Karnouskos S (2012) Service-oriented SCADA and MES supporting petri nets based orchestrated automation systems. In: IECON 2012—38th annual conference on IEEE industrial electronics society, pp 6144–6150. doi:10.1109/IECON.2012.6389076
9. Colombo AW, Karnouskos S (2009) Towards the factory of the future: a service-oriented cross-layer infrastructure. In: ICT shaping the world: a scientific view. European Telecommunications Standards Institute (ETSI), Wiley, New York, pp 65–81
10. Colombo AW, Karnouskos S, Mendes JM (2010) Factory of the future: a service-oriented system of modular, dynamic reconfigurable and collaborative systems. In: Benyoucef L, Grabot B (eds) Artificial intelligence techniques for networked manufacturing enterprises management. Springer, London. ISBN 978-1-84996-118-9
11. Comella-Dorda S, Wallnau K, Seacord R, Robert J (2000) A survey of black-box modernization approaches for information systems. In: Proceedings of the international conference on software maintenance (ICSM'00), IEEE computer society, Washington, DC, USA, p 173. http://dl.acm.org/citation.cfm?id=850948.853443
12. Deugd SD, Carroll R, Kelly K, Millett B, Ricker J (2006) Soda: service oriented device architecture. IEEE Pervasive Comput 5(3):94–96, c3. doi:10.1109/MPRV.2006.59, http://dx.doi.org/10.1109/MPRV.2006.59
13. Diedrich C, Bangemann T (2007) PROFIBUS PA instrumentation technology for the process industry. Oldenbourg Industrieverlag, GmbH. ISBN 13 978-3-8356-3125-0
14. Fitch J, Li H (2010) Challenges of SCADA protocol replacement and use of open communication standards. In: 10th IET international conference on developments in power system protection (DPSP 2010). Managing the change, pp 1–5. doi:10.1049/cp.2010.0220
15. Ganguly J, Vogel G (2006) Process analytical technology (PAT) and scalable automation for bioprocess control and monitoring—a case study. Pharm Eng 26(1):8
16. Großmann D, Braun M, Danzer B, Riedl M (2013) FDI field device integration. VDE, Verlag. ISBN 978-3-8007-3513-6
17. Hérault C, Thomas G, Lalanda P (2005) Mediation and enterprise service bus: a position paper. In: Proceedings of the first international workshop on mediation in semantic web services (MEDIATE), CEUR workshop proceedings. http://www.ceur-ws.org/Vol-168/MEDIATE2005-paper5.pdf
18. Hoffmann P, Maksoud TM, Schumann R, Premier GC (2010) Virtual commissioning of manufacturing systems: a review and new approaches for simplification. In: 24th European conference on modelling and simulation (ECMS 2010), Kuala Lumpur
19. Huovinen M (2010) Large-scale monitoring applications in process industry. Ph.D. thesis. Tampere University of Technology, Tampere. http://dspace.cc.tut.fi/dpub/bitstream/handle/123456789/6512/huovinen.pdf
20. IEC (2006) Function blocks (FB) for process control—part 3: electronic device description language (EDDL) (IEC 61804–3)
21. IEC (2007) Enterprise-control system integration—part 3: activity models of manufacturing operations management (IEC 62264–3)
22. Karasavvas KA, Baldock R, Burger A (2004) Bioinformatics integration and agent technology. J Biomed Inf 37(3):205–219. doi:10.1016/j.jbi.2004.04.003, http://dx.doi.org/10.1016/j.jbi.2004.04.003
23. Karnouskos S, Colombo AW (2011) Architecting the next generation of service-based SCADA/DCS system of systems. In: 37th annual conference of the IEEE industrial electronics society (IECON 2011), Melbourne, Australia
24. Karnouskos S, Baecker O, de Souza LMS, Spiess P (2007) Integration of SOA-ready networked embedded devices in enterprise systems via a cross-layered web service infrastructure. In: Proceedings of 12th IEEE international conference on emerging technologies and factory automation (ETFA 2007), Patras (Best Paper Award), pp 293–300. doi:10.1109/EFTA.2007.4416781, http://www.iestcfa.org/bestpaper/etfa07/2194.pdf

25. Karnouskos S, Bangemann T, Diedrich C (2009) Integration of legacy devices in the future SOA-based factory. In: 13th IFAC symposium on information control problems in manufacturing (INCOM), Moscow, Russia
26. Karnouskos S, Colombo AW, Jammes F, Delsing J, Bangemann T (2010) Towards an architecture for service-oriented process monitoring and control. In: 36th annual conference of the IEEE industrial electronics society (IECON 2010), Phoenix, AZ
27. Karnouskos S, Savio D, Spiess P, Guinard D, Trifa V, Baecker O (2010) Real world service interaction with enterprise systems in dynamic manufacturing environments. In: Artificial intelligence techniques for networked manufacturing enterprises management. Springer, London
28. Karnouskos S, Colombo AW, Bangemann T, Manninen K, Camp R, Tilly M, Stluka P, Jammes F, Delsing J, Eliasson J (2012) A SOA-based architecture for empowering future collaborative cloud-based industrial automation. In: 38th annual conference of the IEEE industrial electronics society (IECON 2012), Montréal, Canada
29. Kasik DJ (2011) The third wave in computer graphics and interactive techniques. IEEE Comput Graphics Appl 31(4):89–93. doi:10.1109/MCG.2011.64
30. Kirkham T, Bepperling A, Colombo AW, McLeod S, Harrison R (2009) A service enabled approach to automation management. In: 13th IFAC symposium on information control problems in manufacturing (INCOM), Moscow, Russia
31. Lindemann L, Thron M, Bangemann T, Grosser O (2011) Integration of medical equipment into SOA—enabling technology for efficient workflow management. In: 2011 IEEE 16th conference on emerging technologies factory automation (ETFA), pp 1–8. doi:10.1109/ETFA.2011.6059233
32. Mahnke W, Leitner SH, Damm M (2009) OPC unified architecture. Springer, Heidelberg. ISBN 978-3-540-68899-0
33. Makris S, Michalos G, Chryssolouris G (2012) Virtual commissioning of an assembly cell with cooperating robots. Adv Decis Sci 2012:11. doi:10.1155/2012/428060, http://dx.doi.org/10.1155/2012/428060
34. Mendes J, Leitao P, Colombo A, Restivo F (2008) Service-oriented control architecture for reconfigurable production systems. In: 6th IEEE international conference on industrial informatics 2008 (INDIN 2008), pp 744–749. doi:10.1109/INDIN.2008.4618201
35. Minor J (2011) Bridging OPC-UA and DPWS for industrial SOA. Master's thesis. Tampere University of Technology, Tampere. http://dspace.cc.tut.fi/dpub/bitstream/handle/123456789/20954/minor.pdf
36. PROFIBUS (2001) FDT interface specification: specification for PROFIBUS device. Technical report, PROFIBUS User Organization. http://www.profibus.com/download/device-integration/
37. Sommer J, Gunreben S, Feller F, Kohn M, Mifdaoui A, Sass D, Scharf J (2010) Ethernet—a survey on its fields of application. IEEE Commun Surv Tutorials 12(2):263–284. doi:10.1109/SURV.2010.021110.00086
38. Taisch M, Colombo AW, Karnouskos S, Cannata A (2009) SOCRADES roadmap: the future of SOA-based factory automation
39. ZVEI (2011) Manufacturing execution systems (MES)—industry specific requirements and solutions. German Electrical and Electronic Manufactures' Association (ZVEI). http://www.zvei.org/en/association/publications/Pages/Manufacturing-Execution-Systems-MES.aspx. ISBN 978-3-939265-23-8

Chapter 3
The IMC-AESOP Architecture for Cloud-Based Industrial Cyber-Physical Systems

Stamatis Karnouskos, Armando W. Colombo, Thomas Bangemann,
Keijo Manninen, Roberto Camp, Marcel Tilly, Marek Sikora,
François Jammes, Jerker Delsing, Jens Eliasson,
Philippe Nappey, Ji Hu and Mario Graf

Abstract A coherent architectural framework is needed to be able to cope with the imposed requirements and realise the vision for the industrial automation domain. Future factories will rely on multi-system interactions and collaborative cross-layer management and automation approaches. The service-oriented architecture paradigm empowered by virtualisation of resources acts as a lighthouse. More specifically by integrating Web services, Internet technologies, Cloud systems and the power of the Internet of Things, we can create a framework that has the possibility of empowering seamless integration and interaction among the heterogeneous stakeholders in the future industrial automation domain. We propose here a service architecture that attempts to cover the basic needs for monitoring, management, data handling, integration, etc., by taking into consideration the disruptive technologies and concepts that could empower future industrial systems.

S. Karnouskos (✉) · J. Hu · M. Graf
SAP, Karlsruhe, Germany
e-mail: stamatis.karnouskos@sap.com

J. Hu
e-mail: ji.hu@sap.com

M. Graf
e-mail: mario.graf@sap.com

A. W. Colombo
Schneider Electric, Marktheidenfeld, Germany
e-mail: armando.colombo@schneider-electric.com

A. W. Colombo
University of Applied Sciences Emden/Leer, Emden, Germany
e-mail: awcolombo@technik-emden.de

T. Bangemann
ifak, Magdeburg, Germany
e-mail: thomas.bangemann@ifak.eu

K. Manninen
Honeywell, Kuopio, Finland
e-mail: keijo.manninen@honeywell.com

A. W. Colombo et al. (eds.), *Industrial Cloud-Based Cyber-Physical Systems*,
DOI: 10.1007/978-3-319-05624-1_3, © Springer International Publishing Switzerland 2014

3.1 Introduction and Vision

The future industrial automation systems are expected to be complex system of systems [4, 8] that will empower a new generation of today hardly realisable applications and services. This will be possible due to several disruptive advances [11], as well as the cross-domain fertilisation of concepts and amalgamation of IT-driven approaches in traditional industrial automation systems. The factory of the future will rely on a large ecosystem of systems where collaboration at large scale [17] will take place. This is only realisable due to the distributed, autonomous, intelligent, pro-active, fault-tolerant, reusable (intelligent) systems, which expose their capabilities, functionalities and structural characteristics as services located in a 'Service Cloud'. Multidisciplinary in nature, the factory appears as a new dynamic cyber-physical infrastructure, which links many component systems of a wide variety of scales, from individual groups of sensors and mechatronic components to e.g. whole control, monitoring and supervisory control systems, performing, e.g. SCADA, DCS and MES functions. The resulting combined systems are able to address problems which the individual components alone would be unable to do, and to yield management, control and automation functionality that is only present as a result of the creation of new, 'emergent' information sources, and a result of cooperation, composition of individual capabilities, aggregation of existing and emergent features [3].

Today, plant automation systems are composed and structured by several domains, viewed and interacting in a hierarchical fashion following mainly the specifications of standard enterprise architectures [19]. However, with the empowerment offered by modern service-oriented architectures, the functionalities of each system or even device can be offered as one or more services of varying complexity, which may

R. Camp
FluidHouse, Jyväskylä, Finland
e-mail: roberto.camp@fluidhouse.fi

M. Tilly
Microsoft, Unterschleißheim, Germany
e-mail: marcel.tilly@microsoft.com

M. Sikora
Honeywell, Prague, Czech Republic
e-mail: marek.sikora@honeywell.com

F. Jammes · P. Nappey
Schneider Electric, Grenoble, France
e-mail: francois2.jammes@schneider-electric.com

P. Nappey
e-mail: philippe.nappey@schneider-electric.com

J. Delsing · J. Eliasson
Luleå University of Technology, Luleå, Sweden
e-mail: jerker.delsing@ltu.se

J. Eliasson
e-mail: jens.eliasson@ltu.se

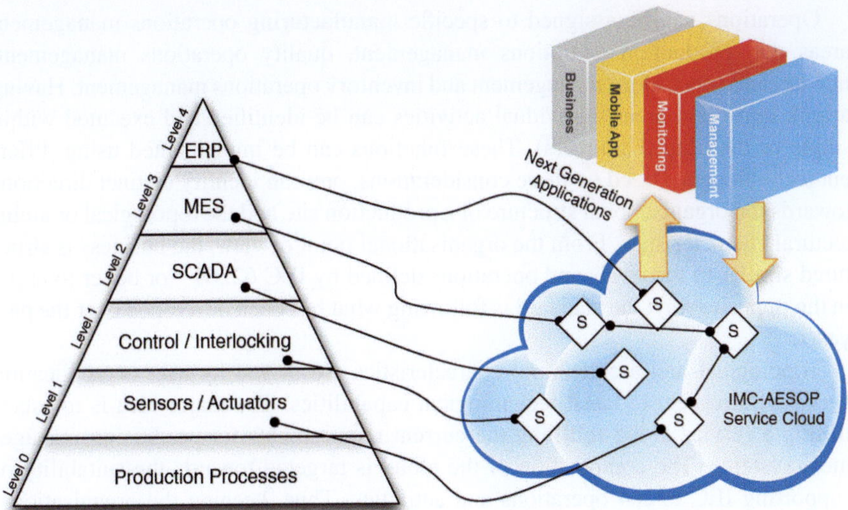

Fig. 3.1 Industrial automation evolution: complementing the traditional ISA-95 automation world view (*pyramid* on the *left side*) with a flat information-based infrastructure for dynamically composable services and applications (*right side*)

be hosted in the cloud and composed by other (potentially cross-layer) services, as depicted in Fig. 3.1. Hence, although the traditional hierarchical view coexists, there is now a flat information-based architecture that depends on a big variety of services exposed by the cyber-physical systems [1], and their composition. Next-generation industrial applications can now rapidly be composed by selecting and combining the new information and capabilities offered (as services in the cloud) to realise their goals. The envisioned transition to the future cloud-based industrial systems is depicted in Fig. 3.1.

Several efforts so far have been directed towards defining structural and architectural aspects of production management systems. The most popular and applied in practice are the definitions set up within the ISA-95/IEC 62264 standard (www.isa-95.com). Typically, today's production systems (factory and process) are structured in a hierarchical way in a five-level hierarchical model. IEC 62264 additionally defines a manufacturing operations management model, implicitly represented by real installations. The standard defines functions mainly associated to levels 3 and 4, objects exchanged, their characteristics and attributes, activities and functions related to the management of a plant, but neither says anything about the implementations (solutions) hosting a specific operation nor the precise assignment to one of the levels 2, 3 or 4. Realisations depend on individual customer needs and the solution provider's strategies. For instance, maintenance management operation may typically be assigned to a Computerised Maintenance Management System (CMMS), a manufacturing execution system (both being typical level 3 solutions), to an Enterprise Resource Planning (ERP) or a distributed control system.

Operations can be assigned to specific manufacturing operations management areas, i.e. production operations management, quality operations management, maintenance operations management and inventory operations management. Having a look into these areas, individual activities can be identified and executed within single or distributed source(s). These functions can be implemented using different technologies. Based on these considerations, one can identify distinct directions towards the organisational structure of a production site and the topological or architectural characteristics. From the organisational point of view, the business is structured similar to the levels and operations defined by IEC 62264—or better to argue in the opposite way: the standard is following what has been developed over the past years.

Topological and architectural characteristics are driven by user or application needs with respect to latest technological capabilities. The major idea is to establishing a service cloud fulfilling the current requirements for production management systems. The composition of the cloud is targeted towards the suitability of supporting IEC 62264 operations and activities. Thus, keeping the organisational aspects established in today's production systems, the migration to future service-based architecture exploiting the capabilities inherent to SOA is approached [5, 6].

In the following sections, considerations taken when designing the new architecture [13, 16] are given to prove the inclusion of the user and application needs. This is the basis for defining services distributed within the cloud or between the cloud and those associated to real physical instances within the architectural framework. Finally, directions of further progress within and behind latest developments are discussed. This architecture is considered as a prelude in realising the vision of cyber-physical systems [1], especially in relation to the 4th Industrial Revolution (referred to as Industrie 4.0 in Germany [9]).

3.2 Design Considerations

In order to design the architecture, a set of use cases and their requirements, as well as concepts and technology trends have been considered. In this section, we focus on the resulting potential directions that may play a key role in the design of the architecture. More specifically, these are:

- Asset Monitoring;
- Backward/Forward Compatibility;
- Creation of Combinable Services and Tools;
- Cross-network Dynamic Discovery;
- Cross-layer Integration and Real-time Interaction;
- Infrastructure Evolution Management;
- Interoperability and Open Exchange Formats;
- System Management;
- Mobility Support;

- Process Monitoring and Control;
- Provision of Infrastructure Services;
- Real-time Information Processing;
- Real-world Business Processes;
- Scalability;
- Service Life Cycle Management;
- System Simulation;
- Unique Asset Identification.

3.2.1 Asset Monitoring

The monitoring of assets is of key importance especially in a highly complex heterogeneous infrastructure. In large-scale systems [17], it will be practically impossible to do effective information acquisition with traditional methods, i.e. often pull the devices for their status. The more promising approach is to have an event-driven infrastructure coupled with service-oriented architectures. As such, any device or system will be able to provide the information it generates (data, alarms, etc.) as an event to the interested entities.

Considering that there exist basically two major kinds of monitoring methods (i.e. feature-based and model-based), the application of the IMC-AESOP approach allows performing both in an individual and in a combined manner. On one side, devices and systems are able to expose feature-based monitoring indexes as services, i.e. monitoring indexes generated by the application of 'relational' functions between sensor signals and information exposed as Web services and the intrinsic characteristics, both structural and behavioural, of the systems and process behind. On the other side, the model-based orchestration approach that is an inherent component of the SOA-based IMC-System (Intelligent Monitoring and Control System) facilitates the creation of new monitoring indexes, this time, model-based monitoring indexes that appear as a result of the composition and orchestration of monitoring services exposed by the orchestrated devices and systems. In this case, the rules to orchestrate or compose the monitoring services, follow the process model functions and are offered as a service usually by a constellation of underlying devices and systems. Emergent behaviours obtained by the orchestration and composition of monitoring services are not a rarity and constitute a clear proof of application of the SoS-paradigm.

Due the close relationship between asset monitoring, control and process monitoring, the components required to create a large-scale system event-driven architecture are mostly the same; the main difference is in that assets extend to anything that can create value for the company, and while this includes the machines that are monitored and controlled, it also extends to personnel, material, energy and to other aspects of the machines used in processes.

3.2.2 Backward/Forward Compatibility

The future industrial infrastructure is expected to be constantly evolving. As such it is important to be (i) backwards compatible in order to avoid breaking existing functionality and (ii) forward compatible, i.e. feature interfaces and interactions as flexible as possible with possible considerations on future functionality and models to come.

While designing components (devices, gateways, mediator) and their architecture one has to consider backward compatibility:

- State of the art and promising emerging technological trends.
- Use of the most used standard technologies (de facto standards in industry) to address a broad range of applications and technical equipment being on the market today.
- Some commonalities can be monitored like concepts for device descriptions or integration mechanisms into SCADA/DCS.
- Step-wise evolutionary process is preferred. Changing too many paradigms, or 'making the step too large and complex', will probably cause acceptance problems within the addressed user community.
- New engineering approaches have to smoothly integrate with the existing ones.
- Taking advantage from past standardisation efforts (e.g. device profiles) will reduce new investments for the establishment of new technology.

While designing components (devices, gateways or mediator) and their architecture one has to consider forward compatibility:

- Focus on most promising open standard technologies.
- Focus on 'living' standards that are actually used, not on 'sleeping' ones that were once defined and never updated or really used in real environments.
- Consideration of hardware capabilities that may have an effect on the architecture and technologies, e.g. single- versus multi-core processor systems, single- or multi-stack architecture, multi-purpose controllers versus single-purpose, etc.
- Software update and download capabilities (ideally with complete lifecycle management).

3.2.3 Creation of Combinable Services and Tools

The trend in software applications is their rapid development by combining existing functionality in a mash-up way. It is expected that this trend will also empower next-generation industrial applications. Since often the development of such functionality is task-oriented, new tools are needed to be developed that ideally can be easily combined in a larger system.

Combinable services and tools should be used. Consider as an example the Unix command line utilities whose functionality can be piped and generate the desirable outcome. Similarly, several tools (proprietary or not) should be combined (orchestrated) and their functionality could provide input to mash-up applications and services. Industrial application development may be greatly eased by following this approach. Typical examples of the design goal here are the functionalities offered by the XML pipeline, i.e. connecting XML processes such as XML transformations and XML validations together, Complex Event Processing (CEP)-driven interactions, service composition, Yahoo! Pipes, etc.

In very large-scale distributed systems, it is desirable to program applications and describe processes at the highest possible level of abstraction. Each service-enabled device abstracts a real piece of equipment functionality or information processing capability in such a way that it can be used as a building block when describing a higher level process. The ability to combine atomic services into higher level composite services, which are themselves abstracted and exposed as services, is one of the fundamental benefits of SOA. A system-of-systems approach, where services can be composed of other services enables creation of mixed systems where high-powered devices, e.g. servers, can provide a complex service composed of a number of underlying services provided by resource-constrained devices. Service orchestration methods, such as BPEL, can be used for managing this.

3.2.4 Cross-Network Dynamic Discovery

Large-scale process control infrastructures typically span multiple sites or multiple subnetworks. For plant operation, maintenance, and engineering, zero-configuration networking can provide the tools for managing a device, or service, throughout its operational lifetime. For instance, a new device or system may automatically announce its presence and allow cross-layer optimisations during its operation. The goal is towards real-time awareness of all cyber-physical parts in the network and their capabilities.

Existing approaches embedded at protocol level, e.g. DHCP or direct IPv6 usage, DPWS, WS-Discovery, etc., or indirect approaches such as network scanning can assist towards identifying connected assets and services. A discovery strategy still needs to be investigated. For instance, when a new device is installed: will the device announce itself or will it rely on a master device scanning the network for changes? Filtering and caching will also have to be considered for large-scale system discovery (e.g. filtering on service type and/or scope …) consisting of heterogeneous networks, e.g. wired and wireless sensor networks, etc.

The use of dynamic device and service discovery, especially over subnetwork boundaries, can create a substantial amount of network traffic. The use of protocols and mechanisms based on polling can have a large impact on a network's performance, which for wireless networks, e.g. WLAN or sensor networks need extra precautions when deploying. Event-based protocols, when combined with caching mechanisms can help mitigate the performance impact from a large-scale deployment of dynamic device and service discovery.

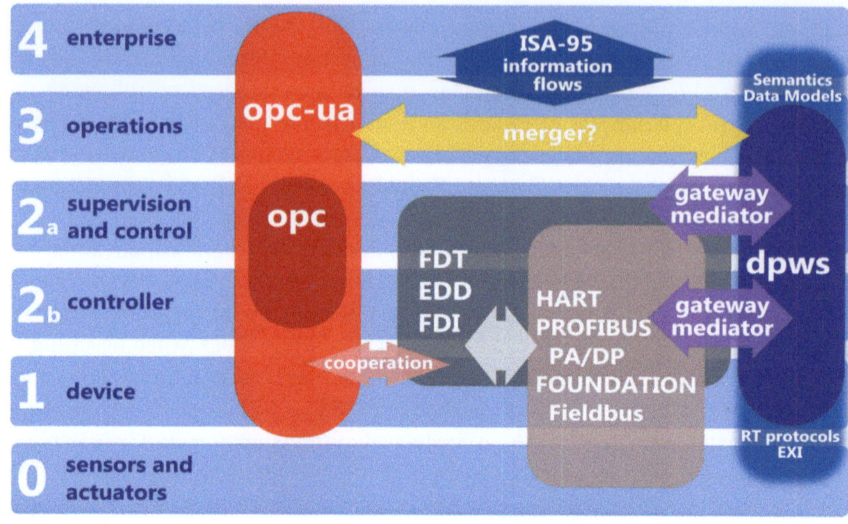

Fig. 3.2 ISA-95 application levels, and relevant current and emerging technologies

Automatic device and service discovery is a key feature for large-scale wireless sensor networks to be maintainable due to the potentially very large number of devices. cross-layer discovery mechanisms help services and systems outside the sensor network to access devices and services inside the networks, thus enabling interoperability and usability. For battery powered devices it is also vital that the discovery protocol of choice is lightweight enough so that the node's energy consumption can be minimised.

3.2.5 Cross-Layer Integration and Real-time Interaction

Cross-layer integration refers to direct communication between different layers in the ISA-95 model (www.isa-95.com) as shown in Fig. 3.2, e.g. a production planning system reading sensors in order to estimate when additional supplies are needed. The aim here is the optimisation at architectural and functional levels of the logical and physical network architecture.

To achieve this, several actions must be taken, e.g. (i) identify activities and information flows, and their performance requirements (hard real-time, soft real-time, right time, etc.), (ii) investigate technologies that can be used to meet the identified performance requirements, (iii) determine standard ways for representing non-functional requirements, such as Quality of Service (QoS), and propose solutions where standards do not exist and (iv) determine optimal network infrastructure patterns, etc.

3.2.6 Infrastructure Evolution Management

Although industrial infrastructures have up to now been designed for the long run, e.g. with 15–20 years lifetime in some cases, in the future they are expected to be more often updated for increased reliability, take advantage of the latest technologies and provision of new functionality. Being technology agnostic of the future advancements, the main challenge is to be able to design today an infrastructure that will be easy to manage and evolve in conjunction with technology.

Better said, the key questions posed are:

- How can one design today the perfect legacy system of tomorrow?
- How can today's functionalities be reused and integrated to tomorrow's infrastructure with minimal effort?
- How can we make sure the transition/migration is smooth and with least impact on key factors such as cost, downtime, maintenance, business continuity, etc.?

A typical example scenario is the automatic software update service on all devices in the network, for security and safety reasons. Another example of the infrastructure evolution is the migration as envisioned in the IMC-AESOP project [5]. It is expected that several migration paths will exist, and each of these paths will additionally have its own number and type of migration steps.

3.2.7 Interoperability and Open Exchange Formats

As next-generation systems will be highly collaborative and will have to share information, interoperability via open communication and standardised data exchange is needed. System engineering of interoperable systems has profound impact on their evolution, migration and future integration with other systems [3, 4, 7]. There are two dimensions of interoperability to be considered: (i) cross-level, i.e. communication between the various levels of the enterprise system, from the plant-floor level up to the enterprise level [19], with systems like ERP or MES; and (ii) cross-domain: the case of multidisciplinary systems where devices and systems of different domains must communicate.

3.2.8 System Management

The next-generation factory systems will be composed of thousands of devices with different hardware and software configurations. There will be a need to automate as much as possible primarily the monitoring part, decision making and also the soft-control of such systems; hence management is of key importance. Management should hide increasing complexity and should provide seamless interaction with the underlying infrastructure such as making it possible to dynamically identify devices, systems and services offered by the infrastructures. It should be possible to do software upgrades and mass reprogramming or reconfiguration of whole systems. Additionally, (remote) visualisation of the real infrastructure is a must, as it will give the opportunity of better understanding and maintaining it.

Management of a heterogeneous network of devices and systems is crucial for the feasibility of a cloud-based large-scale architecture. The use of devices and systems from different manufacturers adds requirements such as flexibility and extensibility to a management system . Using a common communication architecture will mitigate some of these constraints. Scalability and robustness are also important factors when the number of managed (SOA-enabled) devices increases. A management system must be able to effectively support hundreds of thousands of devices with different software and hardware platforms from different vendors.

3.2.9 Mobility Support

In the factory of the future where modern automation systems are in place, the operators are not bound to specialised control centres, but will be able to control and monitor the processes in the shop floor using mobile HMIs. This enables access to real-time measurements and statistics at any time and location. Mobility support also enables monitoring of mobile machinery (automatic loaders, robots, vehicles, etc.).

Mobility will need to be considered towards different angles:

- support for mobile devices, e.g. being used as HMIs,
- support for mobility of devices, i.e. where devices are themselves mobile and the implications of this,
- support for mobile users and interaction with static and mobile infrastructure,
- support for mobility of services, e.g. where services actually migrate among various infrastructures and devices following, e.g. user's profile wishes.

3.2.10 Process Monitoring and Control

Although the topology and structure of processing plants are usually fixed, a challenge is still given by the large size of a typical plant, which may have thousands of actuating, sensing and controlling devices. This makes the design, deployment, management and maintenance of a process monitoring and control system significantly more difficult. An SOA-based approach should address the key challenges to enable maximum system flexibility through its entire life cycle.

Here, one has to consider several megatrends in the process automation industry. For instance, process automation companies are following trends and adopting technologies from the IT sector, such as virtualisation and cloud computing, which are being leveraged in deployments of large-scale process monitoring and control systems/DCS systems. Additionally, exploitation of wireless communication further decreases wiring costs and enables to deploy more devices (sensors, actuators, controllers), and thus, extend the scope and enhance quality of process automation functions.

However, as the automation technology increases in complexity and sophistication, operations professionals are faced with increased volumes of data and information they have to process. In addition, end-users, i.e. industrial operating companies experience reductions in skilled resources. Together with data overload and growing safety, security and environmental concerns, this means that fewer people in operations teams must respond faster, handle more complex processes and make better decisions with bigger consequences.

Process monitoring and control should be eased by the architecture. More specifically, it should be possible to easily decouple the device-specific aspects from the more abstract process ones, and enable the various stakeholders to fulfil their roles. In potentially federated infrastructures, processes may need to be coordinated to avoid side effects that could hamper production lines or avoid intervene with other business goals.

3.2.11 Provision of Infrastructure Services

In the future complex infrastructure envisioned, it cannot be expected that all devices (especially resource constrained ones) and systems will always implement the full stack of software that may assist them in interacting with other systems and their services. As such, auxiliary infrastructure services are needed that will enable collaboration of systems and exchange of information.

Therefore generic services need to be designed and put in place. This implies:

- assumption about generic services hosted at devices and more complex systems,
- generic services provided by the infrastructure itself and assurance that devices and systems can interact with them,
- dynamic discovery of additional (customised) services and easy interaction with them.

As an example the infrastructure services should enable (i) peer-to-peer device/system collaboration (horizontal collaboration) and (ii) device to business collaboration (vertical collaboration).

What is envisioned and wanted is that the infrastructure enables the horizontal and vertical collaboration and integration [14]. Several requirements that would enable easy integration and collaboration have already been identified, especially when this concerns devices in systems. Basically, the infrastructure services should enable collaboration, and therefore we need to consider issues such as dynamic collaboration, extensibility, resource utilisation, description of objects (interface), semantic description capabilities, inheritance/polymorphism, composition/orchestration, pluggability, service discovery, (web) service direct device access, (web) service indirect device access (gateway), brokered access to events, service life cycle management, legacy device integration, historian, device management, security and privacy, service monitoring [15].

3.2.12 Real-time Information Processing

Real-time information processing is a broader topic. We have to distinguish between the technical challenges in hard real-time processing which is about predictive and deterministic behaviour on a device and processing of information with low latency from data sources (e.g. sensors) to the consumer, such as dashboard (or operator in front of dashboard), or database, etc.

For next-generation applications to be able to react timely, apart from real-time information acquisition, we also need real-time information processing. Real-time information processing includes also several other high performance set-ups, e.g. in-memory databases, effective algorithms and even potential collaborative approaches for pre-filtering or pre-processing of information for a specific (business) objective and complex analysis of relevant (stream) events in-network and on-device.

Complex Event Processing (CEP) for processing information in conjunction with CEP functionalities on the edge devices are expected to empower us with new capabilities and an architecture should integrate such concepts. Since CEP relies on several steps such as event-pattern detection, event abstraction, modelling event hierarchies, detecting relationships (such as causality, membership or timing) between events, abstracting event-driven processes, etc., their requirements and design considerations must also be integrated.

3.2.13 Real-World Business Processes

With the standardisation and easier integration of monitoring and control capabilities in higher layers, the new generation of business processes can rely on timely acquired data exchange with the shop floor. This has as a result the potential to enhance and further integrate the real world and its representation in business systems in a real-time manner.

It is expected that the business modellers will be able to design processes that interact with the real world possibly in a service-oriented way [14, 18], and based on the information acquired they can take business relevant decisions and execute them. We consider strong integration with enterprise services, among other things, as well as the tuning of a large-scale system of systems infrastructure to the business objectives [14].

3.2.14 Scalability

Scalability is a key feature for large-scale systems [17]. There are two kinds of scalability:

- *Vertical scalability* (*scale up*). To scale vertically (or scale up) means to add resources to a single node in a system, e.g. add CPUs or memory to a single computer. Such vertical scaling provides more resources for sharing.
- *Horizontal scalability* (*scale out*). To scale horizontally (or scale out) means to add more nodes to a system, such as adding a new computer to a distributed software application, e.g. scaling out from one Web server system to three.

For industrial systems it is expected that scaling-up of resources available on single devices will emerge anyway. As such the impact should be considered, e.g. at SCADA/DCS/PLC, etc., to assess what capabilities can be assumed by large-scale applications, e.g. monitoring. Scaling-out is also a significant option to follow, especially relevant to nodes having attached a large number of devices, e.g. a SCADA system or even a monitoring application running in the cloud with thousands of metering points monitored.

The IMC-AESOP architectural approach following the SOA paradigm on all levels, must support very large heterogeneous networks and their capabilities, e.g. ranging from gigabit networks to low-bandwidth, energy-constrained networked sensors and actuators connected over unreliable wireless links. This also implies that the overall network must be able support cross-network interaction with devices that are completely different in terms of processing power, bandwidth and energy availability. A one-size-fits-all approach is therefore not applicable; instead, the proposed architecture must incorporate mechanisms that can manage different types of devices, systems and networks. Recourse availability, Quality of Service (QoS) and load balancing are a few examples of what the system architecture must be able to monitor and manage.

3.2.15 Service Life Cycle Management

The service life cycle begins at inception (definition) and ends at its retirement (decommissioning or re-purposing). The service life cycle enables service governance across its three stages: requirements and analysis, design and development, and IT operations. As this is going to be a highly complex system of systems, tackling the life cycle management especially of composite (potentially cross-domain) services is challenging. To what extent support needs to rely on the core parts of the architecture and what can be realised as optional extensible add-ons that are domain-specific is a challenge. There are several technologies which already include the key concepts of service life cycle management, e.g. the Open Services Gateway initiative framework (OSGi) and these should be integrated to enable parallel evolution of the various architecture parts.

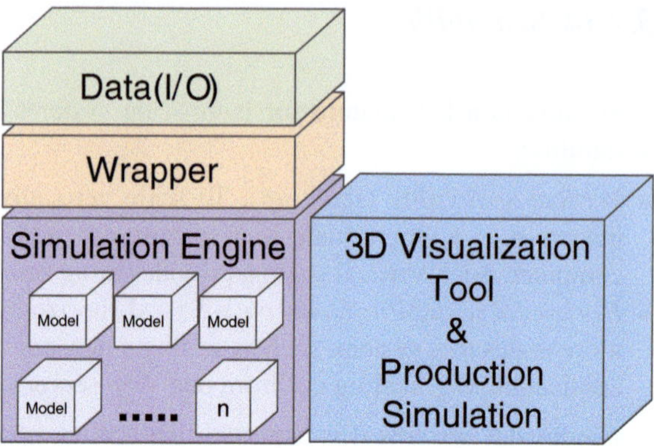

Fig. 3.3 Example of simulation core

3.2.16 System Simulation

Simulations of process systems are pursued in different levels of detail and with different purposes. Three main levels of process simulation to be considered are:

1. *Process design and process control.* At this level the essential operational modes are studied and also the transition between these modes. Main transients and disturbances. Batches and main sequences are analysed. The target is to develop and verify the process design and its control philosophy.
2. *Implementation.* At this level the main focus is on the interface between the field instrumentation and the control system (DCS). There may be less emphasis on the actual process models but more on the signals. The target is to verify the DCS program in terms of logics, interlocks, etc.
3. *Operations.* At this level the ability to operate efficiently is analysed. These simulators can be run in real-time and used as training simulators for the plant operator. The process, the automation system as well as the human interface are represented in the simulator. For aspects related to interoperability and system view, simplifications in the process models and the automation systems may be assumed.

This break-down is quite rough and may significantly overlap; for instance, a simulator for process design and process control design (level 1) can be further developed into a training simulator (level 3) where the actual DCS software is executed (level 2).

One promising architectural approach includes using actual simulation tools and complementing them with an interface/frontend that allows us to simulate actual process and manufacturing systems via an SOA. For example, as shown in Fig. 3.3, having a simulation engine with a message wrapper that can encapsulate simulated events as SOAP messages may allow us to simulate an event-based large-scale System. Different simulation models can be placed inside the simulation engine, each having certain pre-programmed behaviour that can help represent actual devices. It is also possible to complement this architecture with 3D visualisation and production simulation to have a virtually complete system of systems. This kind of architecture approach could allow simulations on levels 2 and 3 mentioned previously. Since the same system could be coupled with SCADA or other supervision systems, user operation/training simulations can be performed in parallel with implementation tests.

Industrial process plants can be considered as complex systems, where a change in one subprocess may result in unexpected consequences in other parts of the plant. Nevertheless, autonomicity of the subprocesses and the subsystems is needed to achieve overall evolution. Therefore, a holistic system analysis is needed to identify possible conflicts and side effects at an early stage. Simulations of process systems is pursued at different levels with varying detail. It is expected that system-wide simulations will assist in designing, building and operating future industrial infrastructure and their interactions.

3.2.17 Unique Asset Identification

Some kind of standardised universal asset identification and addressing mechanism is required for the architecture to be able to support service-oriented targeted communication between and inside the systems. This addressing mechanism should be flexible, scalable and should not introduce additional overhead in configuration, performance and complexity.

In the era of Internet of Things, it must be possible to uniquely identify items and their services. Promising approaches include UUID (e.g. combination of unique data such as IP/MAC/serial number, etc.). With IPv6 it might be possible to have these devices directly addressable. Assets are treated as a general case of devices, systems, people and other resources. These assets carry a unique RFID with them, as this tag cannot carry too much information; software is used to link additional information to that asset. Then, other relevant wireless technologies (e.g. ZigBee) may be used to obtain the information from the ID.

Unique asset identification is very closely linked to the monitoring of assets as it enables part of that monitoring. It can be from the location aspect or from the simple awareness of the qualities and properties of the asset that have to be identified. Additionally, the unique device authentication is expected to improve safety (in part by helping to identify counterfeit products and by improving the ability of staff to distinguish between devices that are similar in appearance but serve different functions). It would be useful if this is coupled with dynamic discovery. In a typical scenario a new device is plugged in the network by plant maintenance staff, and it is dynamically discovered and registered. Subsequently, an operator is presented in his (mobile) tablet the new device and assigns the necessary configuration to it (this may be done remotely). Hence, it is important to be able to distinguish and target specific devices even remotely.

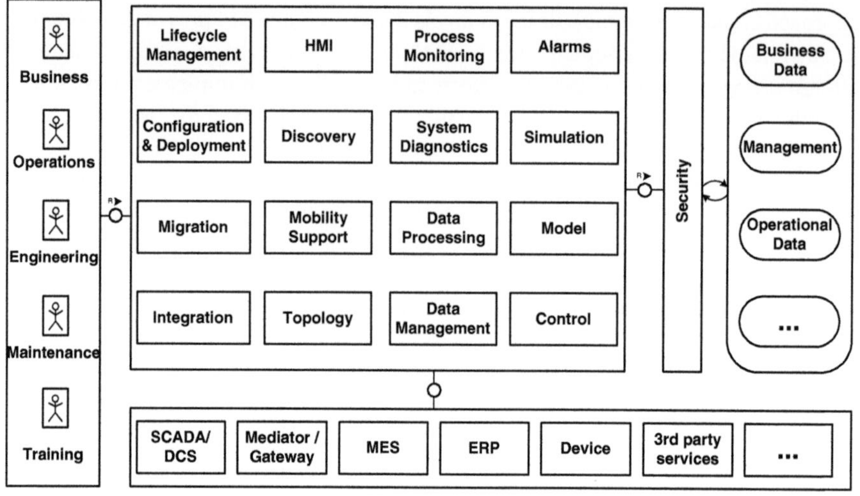

Fig. 3.4 Architecture overview

3.3 A Service-Based Architecture

The IMC-AESOP project follows a service-oriented architecture, a general overview of which is depicted at high level in FMC notation (www.fmc-modeling.org) in Fig. 3.4. On the left side we see the users who interact with the services (depicted in the middle). The data depicted on the far right side can be accessed with the necessary credentials. Although we consider that the majority of these services will run on the 'Cloud' some of these may be distributed and run in more lightweight versions on devices or other systems. As long as the SOA-based interaction is in place, they are considered as part of the general architecture view.

3.3.1 User Roles

Several *'user roles'* are envisioned to interact with the architecture either directly or indirectly as part of their participation in a process plant. The roles define actions performed by staff and management.

The *Business* role handles overall plant management and administration and ensures long-term effectiveness and strategic planning. From an IT point of view, this role is operating in the enterprise layer of a process plant, interacting with supporting systems such as Enterprise Resource Planning (ERP), Enterprise Asset Management (EAM), Operational Risk Management (ORM), etc.

The *Operations* role performs the normal daily operation of the plant, hence it handles optimisation of the monitor and control processes. It is also responsible for meeting the production targets while ensuring that the plant is running in the most efficient, safe and reliable modes. The tasks performed as part of this role are located at the operations layer and use supporting systems such as Operations Control System (OCS) for monitoring and control of the process infrastructure and process optimisation systems.

The *Engineering* role is here divided into two categories: process engineering and system engineering. The Process Engineer ensures proper design, review, control, implementation and documentation of the plant processes. It also designs the layout of the process and performs optimisation work with Operations. The System Engineer deals with the deployment of new automation devices, software components and machines, manages configurations, infrastructure and networks.

The *Maintenance* role is responsible for the system operation with optimum performance, and ensures that the plant's systems and equipment are in a safe, reliable and fully functional state. The maintenance operations are also part of the operations IT layer of the process plant. The systems that are supporting the tasks performed within the maintenance role are Risk-Based Inspections (RBI) systems, systems monitoring, diagnostics and control, etc.

Training ensures that all plant personnel have a basic understanding of their responsibilities as well as safe work practices. Training is performed on a regular basis by all other roles in order to improve work skills. The training planning for each employee must be harmonised with the management strategy planning and can be performed on-site but also using simulation training systems.

3.3.2 Service Group Overview

As depicted in Fig. 3.4, it is possible to distinguish several service groups, namely: Alarms, Configuration and Deployment, Control, Data Management, Data Processing, Discovery, HMI, Integration, Life Cycle Management, Migration, Mobility Support, Model, Process Monitoring, Security, Simulation, System Diagnostic, Topology. These groups indicate high-level constellations of more fine-grained services. IMC-AESOP has defined some initial services which are listed in detail in Table 3.1.

All of the services are considered essential for next-generation cloud-based collaborative automation systems. Table 3.1 depicts a first prioritisation according to what we consider necessary for future systems. The groups of services have been rated with high priority (+) if they constitute a critical service absolutely mandatory, with medium priority (0) if this is not a critical but nevertheless highly needed service and lastly with low priority (−), which mainly means 'nice to have' services that enhance functionalities but are optional.

While most of these correspond to specific real-world scenarios we consider, expanding the potential scenarios may lead to adjustments on the architecture as such. Within the IMC-AESOP project, several of these have been implemented as proof of concept. There are also several functional requirements which will need to be further evaluated and may depend on domain-specific scenarios. To what extent they might impact the proposed approach is avenue for further research.

Table 3.1 Detailed architecture services and prioritisation

Service group	Services	Priority
Alarms	Alarm configuration	+
	Alarm and event processing	+
Configuration and deployment	Configuration repository	+
	System configuration service	+
	Configuration service	+
Control	Control execution engine	+
Data management	Sensory data acquisition	+
	Actuator output	+
	Data consistency	0
	Event broker	+
	Historian	0
Data processing	Filtering	+
	Calculation engine	0
	Complex event processing service	+
Discovery	Discovery service	+
	Service registry	+
HMI	Graphics presentation	+
Integration	Business process management and Execution service	0
	Composition service	+
	Gateway	+
	Service mediator	+
	Model mapping service	+
	Service registry	+
Lifecycle management	Code repository	−
	Lifecycle management	+
Migration	Infrastructure migration solver	−
	Migration execution service	−
Mobility support	Mobile service management	0
Model	Model repository service	0
	Model management service	0
Process monitoring	Monitoring	+
Security	Security policy management	+
	Security management	+
Simulation	Constraint evaluation	0
	Simulation execution	0
	Simulation scenario manager	0
	Process simulation service	0
System diagnostic	Asset monitor	+
	Asset diagnostics management	+
Topology	Naming service	+
	Location service	+

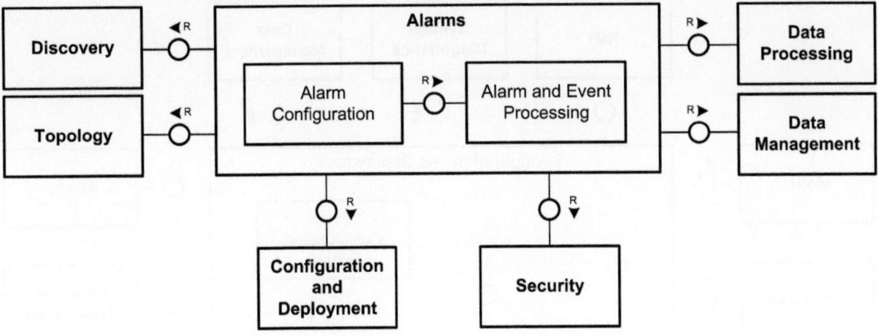

Fig. 3.5 Service group: alarms overview

3.3.3 Alarms

The alarms service group (depicted in Fig. 3.5), contains services for alarm processing and configuration. These services support simple events and complex events that are aggregated from several events. Some of the alarms are generated in lower level services and devices but alarms can be generated also in the alarm processing service using process values and limits. The alarm configuration and processing services also support very flexible hierarchical alarm area definitions.

The alarm configuration service provides help for alarm definitions and maintenance of simple alarms and complex alarms (and events). Each alarm or event can be defined for one or many alarm areas. The alarm areas are hierarchical and there can be several parallel alarm hierarchies. One alarm can belong to one or many alarm hierarchies but it is included only once in one alarm hierarchy.

Complex events are events which are aggregated from several events. They can also use other complex events but the hierarchy is not limited to levels, i.e. one complex event can use complex events (or events) from any level. Complex events are independent of the area definitions but each complex event can belong to one or many alarm area hierarchies. This service is limited to predefined complex events, so modelling of event hierarchies or detecting relationships is not part of it.

The processing service is able to handle thousands of events and map them to alarms coming from different devices in order to filter and aggregate the alarms. The service is based on Complex Event Processing (CEP) principles but it also supports simple (traditional) alarms. It receives the alarm area configuration, simple event configuration and complex event configuration and uses it to process the incoming alarms and events. The service is activated every time a new alarm or event arrives but it can be activated also when the complex event configuration contains time-based activations. The configuration can be hierarchical and complex events can trigger higher level complex events. The complex event processing is typically triggered by an event which was created by another service or application but it can also create its own events, e.g. when configured to monitor some values against the limits.

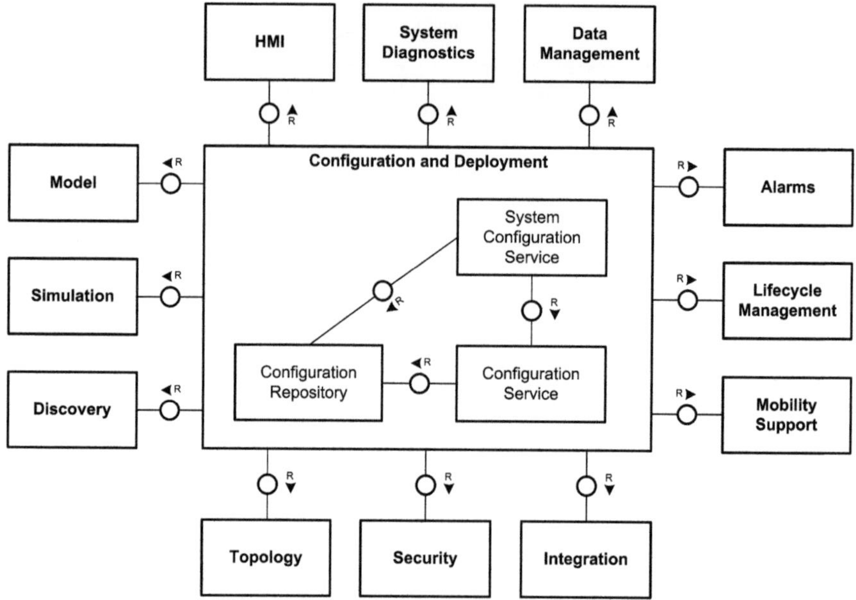

Fig. 3.6 Service group: configuration and deployment overview

Some of the typical alarm area hierarchies are process areas, instrument areas, safety areas, energy areas and quality areas. The plant personnel scope of responsibilities is linked to these area hierarchies.

3.3.4 Configuration and Deployment

The configuration and deployment service group (depicted in Fig. 3.6) is responsible for managing configuration and deployment of various systems from processes to devices. The service group consists of configurable services which enforce and execute the configuration on the device level, system configuration services which deploy configurations of the processes, and a configuration repository where various configuration modes are persisted and retrieved.

The configuration service is needed for the plant control strategy configuration. It uses directly the model service, which supports all the functionalities needed for hierarchical control strategy configuration. In an example scenario where an engineer wants to add a control loop, he would have to add a node, e.g. by sending a POST to https://imc-aesop.eu/configuration and pass all necessary parameters, e.g. node info, attributes, parameters, control algorithms, etc.

The configuration repository service utilises the model repository. It typically has process models for simulation purposes. However, the model repository is not limited

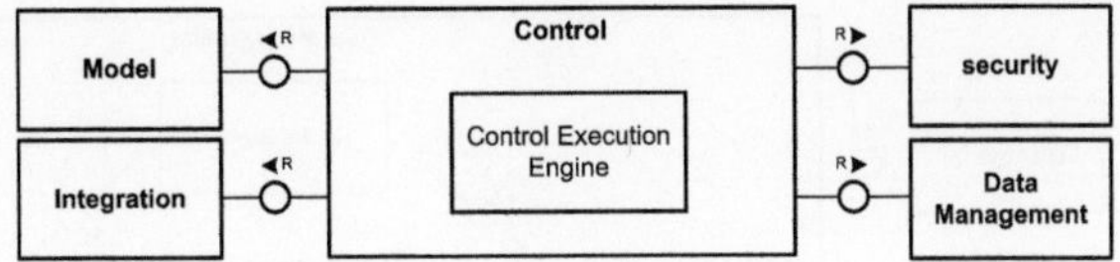

Fig. 3.7 Service group: control overview

to any specific type of hierarchical models, e.g. the configuration repository service utilises the model's structure to save the hierarchical configuration structure. Support for several parallel hierarchical models should be there, e.g. the possibility to add nodes to each hierarchy separately and it is also possible to merge two hierarchies together. Each node contains some kind of process model or information about the process but the model repository service does not understand or care about the internal structure of each node.

The system configuration service provides functionalities to manage configurations for different systems such as processes, SCADA/DCS, PLC and devices. This service is able to check configuration consistency, to send or re-send configuration files to devices, to manage versioned platform-specific implementation of services and to instantiate plant metamodels.

3.3.5 Control

The control service group (depicted in Fig. 3.7), contains the control execution engine service, which is able to execute the process automation configuration or process models. The execution engine services are distributed to several physical nodes and some of these can be redundant. It also supports the typical online (and on-the-fly) changes in configuration while the process is running.

The control execution engine service contains the execution engine that is capable of executing the code generated by the configuration service or the model management service. The executable nodes created by the configuration service typically contain functionalities to control the actual process while the nodes coming from the model management service contain process models. The control execution engine service does not distinguish between these two node types. The requirement for each node is that it must contain the Execute method and follow predefined attributes, e.g. CycleTimeMS, Phase, Priority and ExecutionOrder.

The control execution engine is a distributed service that can run on tens of nodes simultaneously. Some of the nodes are real-time nodes where the deterministic execution is guaranteed. Two or more control execution engines can be combined as a single redundant execution engine. In this case all the redundant execution engines contain exactly the same (configuration or model) nodes but only one engine (at the time) is responsible for the execution and data are copied to the passive engine(s). This responsibility is transferred to another redundant execution engine in case of a hardware failure.

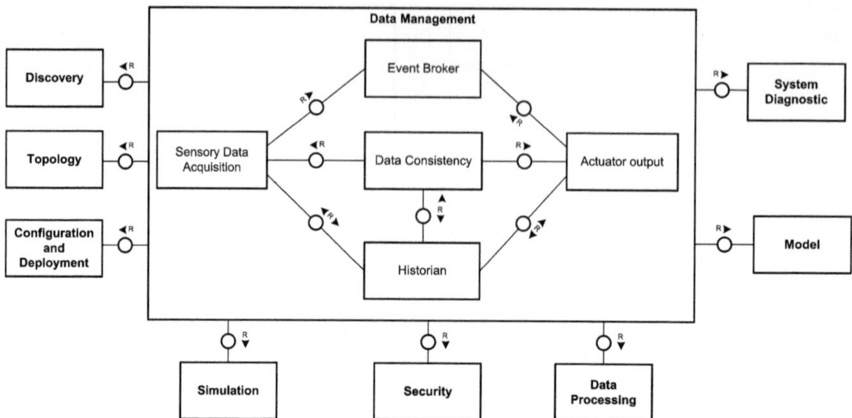

Fig. 3.8 Service group: data management overview

The executable nodes are transferred from the tools when the engineer selects to load the node to the specified execution engine. The execution engine allocates the required memory for the node and adds it to the execution list with the specified cycle time, phase, priority and execution order. The engineer is then able to start the node execution and the execution engine will call the execute method in a specified cycle or when an execution event is received. It is possible to replace the node with a new version online by manually stopping the execution, loading the new version and restarting the execution or 'on-the-fly' by replacing the old version between two execution cycles without losing any control cycles.

3.3.6 Data Management

The data management service group (depicted in Fig. 3.8) encapsulates the functionality of data retrieval, consistency checking, storage and basic eventing. Data management provides services for acquiring data from sensors, consistency checking and plausibility checks, data logging and searching, event generation and actuator control.

The sensory data acquisition service provides an interface for retrieving sensor data. It connects physical devices producing data with higher layer services, such as filtering, eventing and processing, in the architecture. The main function provided by this service is reading of sensor data. It provides methods for typing of the data and mapping to a data model/ontology. Configuration and other features are handled by other services.

The actuator output service is used to control the output of the actuator devices. Typed data, performed by the consistency check service is used to control the physical output of a device's actuator(s). The main function provided by this service is setting and reading of actuator outputs. It provides methods for typing of the data and mapping to a data model/ontology.

The data consistency service validates that data delivered from a device are consistent according to specific rules. The validation can be performed on a device/resource, or within the cloud. Cloud-based validation enables complex queries involving multiple sources of data to be executed. The data consistency service also provides methods for filtering and detection of data that have anomalies. This service allows configuration and querying of consistency rules. Moreover, it provides a way to retrieve the inconsistent data for debugging purposes.

Firing and receiving events constitutes a core concept within the ICM-AESOP architecture, and hence the event broker plays a pivotal role. In general, each service can act either as a producer of events and/or as a consumer of events. As a producer, a service would have to enable consumers to subscribe to topics of events to enable the producer to push events to a given endpoint. A consumer has to provide an endpoint to which a producer can push the events. The event broker is a service that can fire and receive events. The service can subscribe to various topics of event provider and offers an interface for consumers to subscribe to topics of events. In addition, this service uses the historian service to log events for reliability purposes. The event broker service can be used in situations when n producers and m consumers need to be connected. Thus, instead of having $n * m$ registrations the event service reduces the numbers of registrations to $n + m$. Nevertheless, an event service can become also a bottleneck. Therefore, the architecture does not limit the number of event services in a system.

The role of the historian service is to keep and manage a record of a time series of data or events. Historical data can include sensor values, device states, calculated or aggregated values, and diagnostic data. Events of interest can include alarms, state changes, operator instructions, system triggers or any other notification. The data historian exposes an interface for storing, configuring, browsing, updating, deleting and querying historical data and historical events. A typical scenario would be the logging performance data for system diagnostics. The system diagnostics tools benefit from having a view of the state of certain system parameters over time when diagnosing the source of some fault. Similarly, historic data for relevant alarms and events can be kept. Similarly in process optimisation, when trying to optimise a process based on some criteria, historical process data can be used to identify where adjustments can be made.

3.3.7 Data Processing

The data processing service group (depicted in Fig. 3.9), provides services from simple filtering up to complex analytics. This is meant in a functional grouping and is intended to be used on all levels, from device up to the cloud.

Complex event processing is a technology for low-latency filtering, correlating, aggregating and computing on real-world event data. A service offering CEP

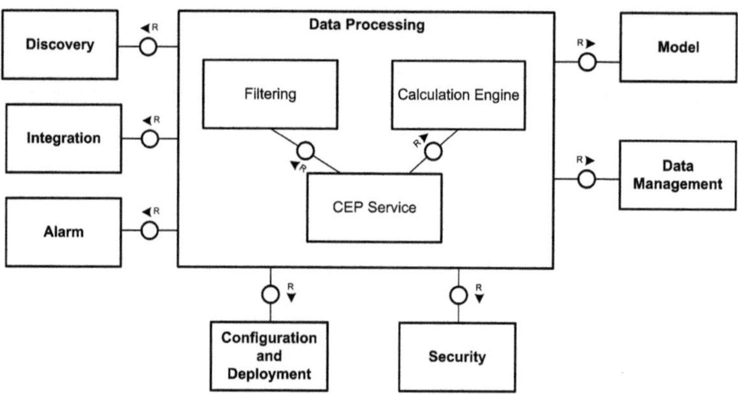

Fig. 3.9 Service group: data processing overview

capabilities enables on one side the consumption of events as inputs and produces (complex) events on the output side. In addition, the service enables the deployment and management of rules (or queries) over the incoming events. These rules (or queries) produce the events on the output. Thus, the service offers also a management API to create, update or delete these rules. The CEP service provides functionality of a complex event processing engine as a service.

The purpose of this calculation engine service is to provide environment for user-defined calculations including numeric and logic operations. The user-defined calculations are additional, perhaps temporary, calculations which are used, e.g. for reporting purposes, process studies, etc. More permanent calculations should be done using normal DCS configurations tools. The user-defined calculations can use, combine and manipulate any process values available in the IMC-AESOP system address space.

3.3.8 Discovery

The discovery service group (depicted in Fig. 3.10), mainly includes services targeting dynamic discovery that allows to find devices/systems/services by type; and location and a registry type service, relying on a known registry endpoint address, featuring at least register, de-register, search and rating of services operations.

Any service, either provided by a physical device within the plant premises or hosted in the cloud, will announce and describe itself when entering the cloud of services. Any other device or service may request more detailed information (service description) or search for available services in the cloud of services. Experimentation in IMC-AESOP demonstrators did show that this discovery mechanism could be combined with a static service enumeration, ensuring that services required for the proper operation of the application have been discovered at runtime.

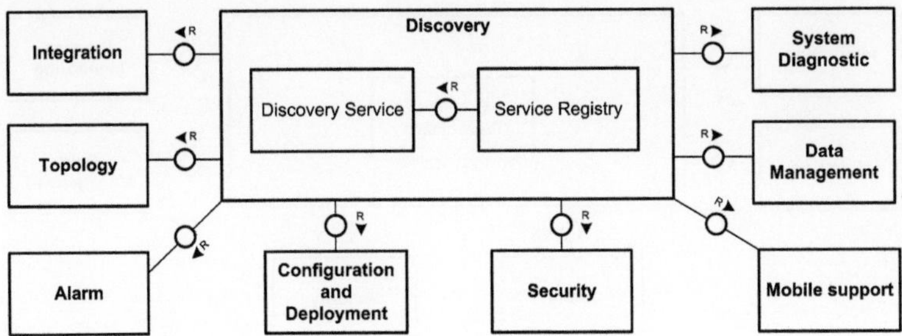

Fig. 3.10 Service group: discovery overview

A typical scenario would be the automatic plug and play. As soon as a device is plugged into the cloud of services, it can automatically search for services that it requires to provide its function and start when these services are available. In the same train of thought, as soon as a device is plugged into the cloud of services, its provided services are automatically registered to provide any management functionality of the cloud of services.

An automatic discovery mechanism, relying on multicasting or broadcasting as described above, is not compatible with all network architectures and all types of services. A service registry is more generally required for SOA-based architectures where services can be hosted both locally and remotely. It is also required for types of services that do not support discovery mechanisms, REST services for instance.

The registry service is used as a repository for all available services across IMC-AESOP architecture. This repository is accessed either (i) by systems and/or devices that register or de-register their services into the registry, mainly at initialisation time; (ii) by systems and/or devices looking for a specific service.

For example, in case the local network is segmented by routers (physical segmentation) or VLANs (virtual segmentation) then both multicast and broadcast communication will be limited to a local subnet and will not spawn multiple network segments. It is therefore useful to consider a discovery proxy mechanism that any endpoint in the system can access, either for registration or for query, independently from its physical location in the network. This discovery proxy service is by essence a service registry. When for instance a new device is connected to the local network and exposes a well-known maintenance service including various device configuration and monitoring methods, it automatically registers its service(s) into the IMC-AESOP services registry. Any monitoring application, looking for maintenance services, can query the registry and retrieve the new device maintenance service endpoint.

3.3.9 HMI

The HMI service group (depicted in Fig. 3.11), contains the graphic presentation service which supports the graphical tools in generic web-based user interface

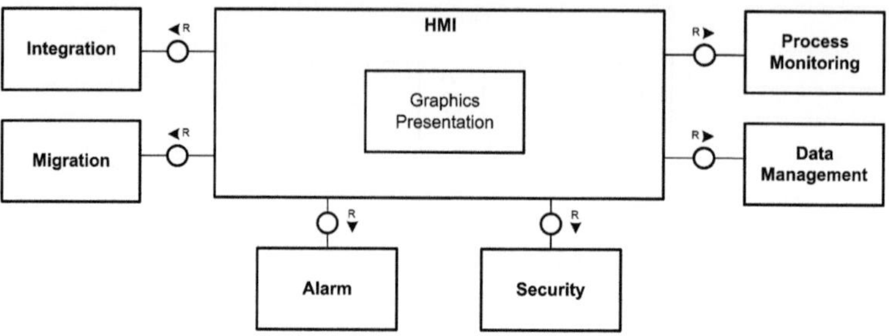

Fig. 3.11 Service group: HMI overview

framework. It provides the generic menu and help functionalities and also the application area where the actual graphical tools are shown.

The graphics presentation service aims at easing interaction with the multiple heterogeneous visualisation devices and applications we expect to populate future systems. We consider a very important and challenging task to design a new framework that will have basic services that will offer the capability to compose graphical user interfaces in a service-driven way. Here the guidelines and concepts from W3C should be followed for the sake of interoperability and openness.

A very simple (and probably only as an intermediate solution) would be to provide each graphical element as a result of a service that could be combined in the screen and utilised by a specific technology. The amount of active content on the pages is minimised, but in some areas it is required because of the performance requirements. However, the active content is transparent for the user and does not require any visible installation procedures or registrations which makes it possible to use various end-devices over the network.

3.3.10 Integration

The integration service group (depicted in Fig. 3.12), enables the combination of functionality for added value. Heterogeneous components with different communication protocols and data models require services to facilitate their interoperable interaction. Business process management and execution, composition, functionality wrappers (gateway and mediator) and model mapping services are part of this service group.

The business process management and execution service manages and executes business processes. The platform exposes processes as higher level services, possibly in the form of a WSDL document with semantic descriptions, and can provide additional tools for controlling and analysing the process.

A composition service provides a platform for managing execution of service compositions. This service would receive as input descriptions of service

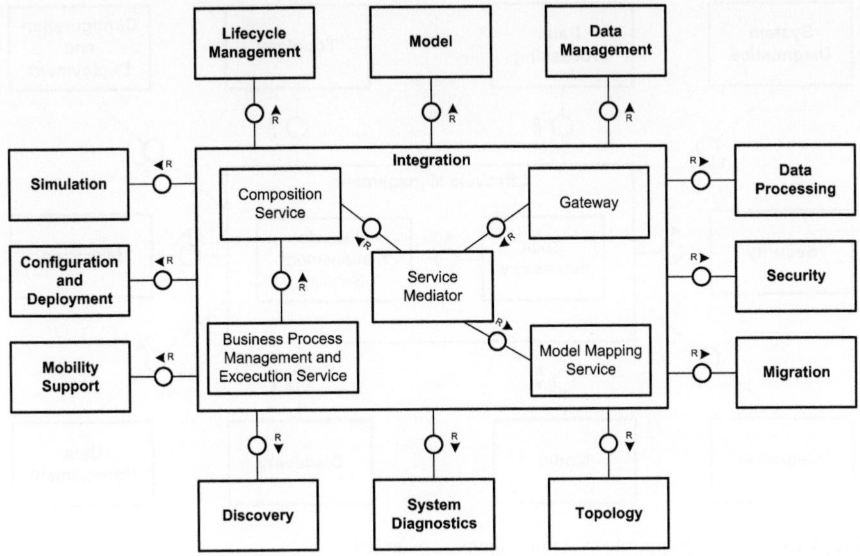

Fig. 3.12 Service group: integration overview

compositions, defined as combinations or sequences of finer grained services along with descriptions of input and output message exchange patterns, logic describing process flow and error conditions. The platform would then expose the business process as an interface to a higher level service, handling any input and output parameters specified. An example would be wrapping often-repeated service invocation patterns as a coarser grained service. An engineer identifies a service invocation pattern that is often repeated in higher level business processes, for example a multi-step start-up or shutdown sequence, or a complex heating or cooling cycle. He then describes this pattern using supported notation, and exposes it as a coarse-grained service with more business relevance.

In the context of the integration of legacy systems and devices in the plant, Gateways and Mediators are used to expose the legacy data as high-level services using the state-of-the-art meta-model. The model mapping service encapsulates the conversion between the legacy and IMC-AESOP data models, including the semantic level. The model mapping service is used both by Mediators and Gateways. The legacy models and mapping rules are typically initialised during the start-up of the system but they can be updated all along the plant life cycle. The mapping service is called either by the Gateways, generally to react to specific demands, or by mediators on a regular basis.

A gateway service provides the means to encapsulate legacy protocol and application objects logic. Encapsulation is supported by the model mapping service. It is used to introduce a non-standard service contract with high-technology coupling. Based on the model mapping, this service represents semantics of legacy components as they expose within the legacy system.

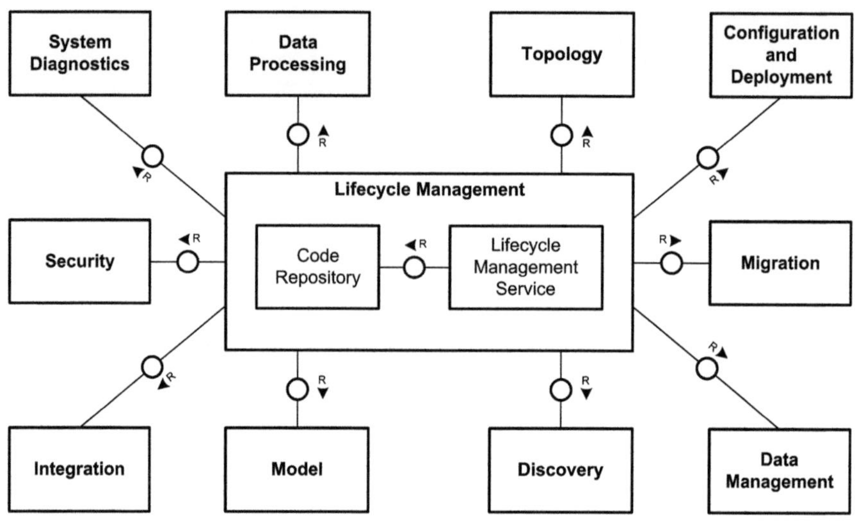

Fig. 3.13 Service group: life cycle management overview

A service mediator service provides the means to encapsulate legacy protocol and application objects logic of a single or several legacy components data and associated functions. Encapsulation is supported by the model mapping service, the business process management and execution service and the gateway service. It is used to introduce a non-standard service contract with high-technology coupling. Based on the model mapping, this service uses data retrieved from legacy components to provide enhanced semantics, legacy components are not able to expose by themselves.

3.3.11 Life Cycle Management

The life cycle management service group (depicted in Fig. 3.13) is a crucial system dealing with the management and evolution of the infrastructure itself. The services provided cover system life cycle aspects such as maintenance policies, versioning, service management and also concepts around staging (e. g. test, validation, simulation, production).

Services will need to be maintained, (re)deployed, upgraded, etc., over a longer period of time. Hence it makes sense to have code repositories that maintain the various implementations and potentially also the source code of these (if available). The code repository should cover needs such as:

- The need to find code based on criteria, e.g. author, execution environment, platform, technology, description, performance, etc.
- The need to describe the developed code based on widely acceptable templates and vocabulary. Using these, the vision of semantic web is promoted and also

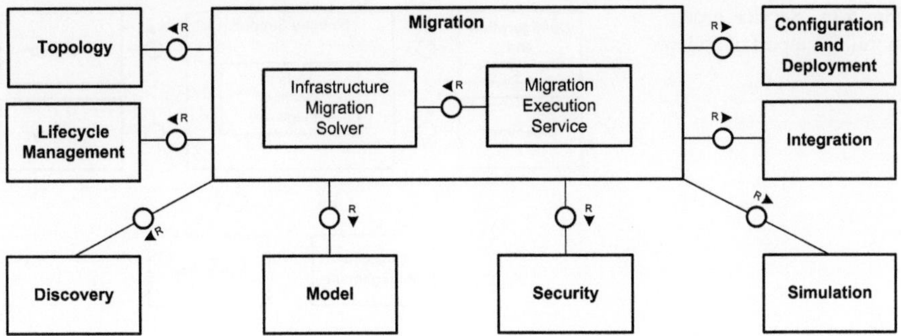

Fig. 3.14 Service group: migration overview

the automation of tasks such as search, management, etc., can be delegated to intelligent technologies, e.g. intelligent mobile agents.

- The need to integrate security, trust and availability from day one.

Life Cycle Management is to support the various services envisioned in IMC-AESOP from an infrastructure point of view. This indicates enabling support for key aspects including deployment, migration and discovery.

3.3.12 Migration

The migration service group (depicted in Fig. 3.14), provides support to migrate a legacy system to a new SOA-based system. This group contains two main services, the infrastructure migration solver and the migration execution service. The infrastructure migration solver helps identify dependencies and offers migration strategies and instructions. The migration execution service implements migration process according to dependencies and instructions.

Under the provision of a set of constraints and a model this service evaluates the feasibility of solving a potential migration from the current landscape to the new one. This is a complex process, where the details are to be captured on the model and constraints themselves. The migration execution service executes the changes needed as identified by the migration solver service. It is assumed that this may be a workflow and step-by-step process where hardware and software parts are migrated.

3.3.13 Mobility Support

The mobility support service group (depicted in Fig. 3.15), provides services for managing mobile assets, such as mapping/changing IP addresses, asset locations, tracking, etc. It also provides data synchronisation services to enable up-to-date data access and sharing for mobile services and devices.

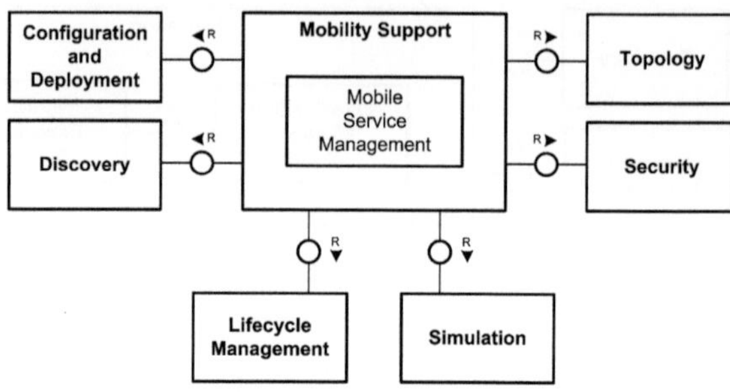

Fig. 3.15 Service group: mobility support overview

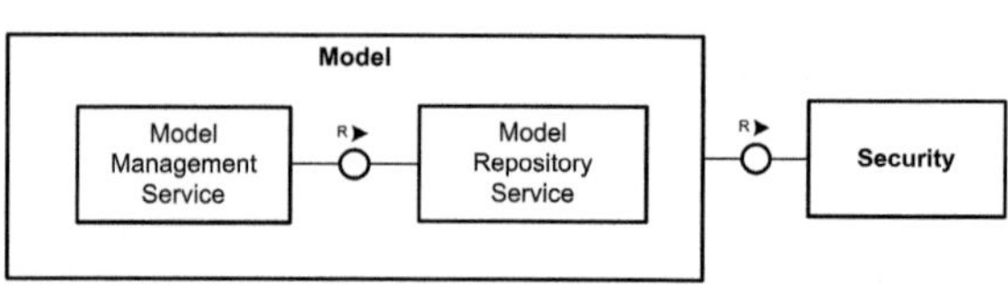

Fig. 3.16 Service group: model overview

3.3.14 Model

The model service group (depicted in Fig. 3.16), contains services for model management and repository. These services are generic and can be used for process automation configurations or process models, and are not limited to these hierarchical models. The model repository takes care of the structure but not the content, hence it is able to support several model types.

Plant information model is a model which contains, e.g. hierarchical plant control strategy or a hierarchical plant process model. However, the model can be created for any purpose and is not limited to these examples. Potentially plant maintenance/service, production optimisation or multivariable controls might require a different kind of hierarchical plant information model.

The hierarchy is maintained by the model repository service which is agnostic to the actual structure of each node. The model management service links the nodes, parameters, attributes and methods together. After these definitions it is possible to execute the hierarchy (or part of it) on a distributed or centralised execution engine. The model management service also contains some predefined basic data types (e.g. float, double, int, unsigned int, byte, string, etc.), some predefined enumeration types, as well as some attributes (e.g. CycleTimeMS, Phase, Priority, ExecutionOrder, etc.). However, with the model management service it is also possible to add new data types, structures, enumerations and attributes.

The model repository service provides an interface to model repository. These models are typically process models for simulation purpose. However, the model repository is not limited to any specific type of hierarchical models. The service supports several parallel hierarchical models. It is possible to add nodes to each

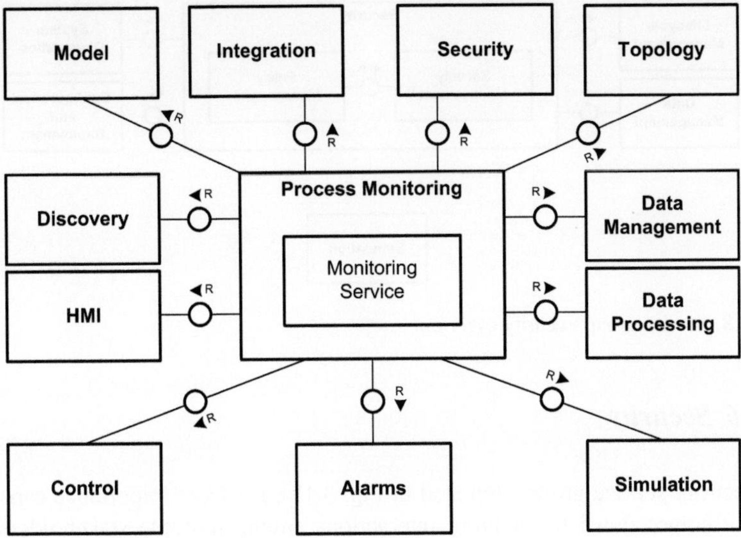

Fig. 3.17 Service group: process monitoring overview

hierarchy separately and it is also possible to merge two hierarchies together. Each node contains some kind of process model or information about the process but the Model Repository service does not understand or care about the internal structure of each node.

3.3.15 Process Monitoring

The process monitoring service group (depicted in Fig. 3.17), serves as the entry point for the operator through the HMI. It is used to gather information relevant to the physical process, e.g. adding semantics to the raw sensor data gathered from data processing and data management service groups. It also deals with process-related alarms and events.

This service provides an interface to collect and analyse process data using capabilities of other architectural components, compare data against expected or simulated results, attach process semantics to raw data and calculate process-related KPIs. An example would be the operator monitoring the relevant parameters or measurements related to the physical process, including levels, flow rates, temperatures, etc. These values, or calculations and aggregates of these values, can be displayed on an HMI.

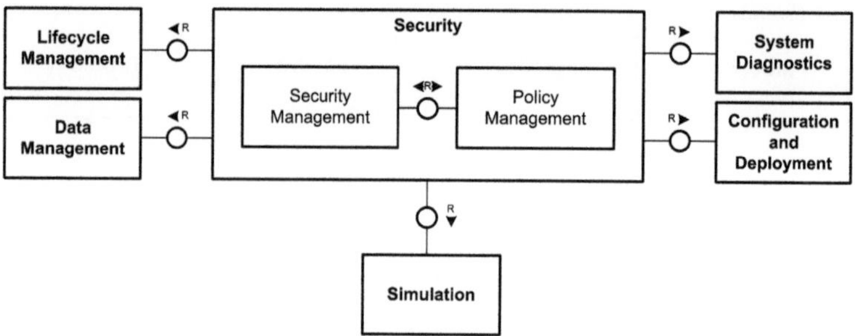

Fig. 3.18 Service group: security overview

3.3.16 Security

The security service group (depicted in Fig. 3.18), is of key importance especially when it comes down to enabling interactions among multiple stakeholders with various goals and access levels. The security management focuses on enforcement or execution of security measures and policy management is about definition and management of security rules or policies. The security services are implicitly used by all architecture services.

In IMC-AESOP, services play a central role which connects heterogeneous devices with monitoring and control applications and makes diverse service applications and business processes interoperable. Therefore, the security architecture of IMC-AESOP mainly focuses on service-related security components such as security management and policy management. Security management focuses on enforcement or execution of security measures and policy management is about definition and management of security rules or policies.

The security management service provides fundamental security functionalities such as authentication, authorisation, confidentiality, digital signatures, etc. The service is also able to provide deployment and enforcement support for security policies and rules defined by security administrators. The duties of the security policy management service are twofold: (i) manage the policies which define access rights to devices or services depending on the user type (identity based or role-based) and (ii) manage the policies which define identity federation to establish federation among various service domains.

3.3.17 Simulation

The simulation service group (depicted in Fig. 3.19), is practically related to every other service group in the architecture as it aims at simulation of multiple systems and their processes. It is in charge of evaluating constraints and simulating execution. It also manages simulation scenarios and uses the exposed simulation endpoints

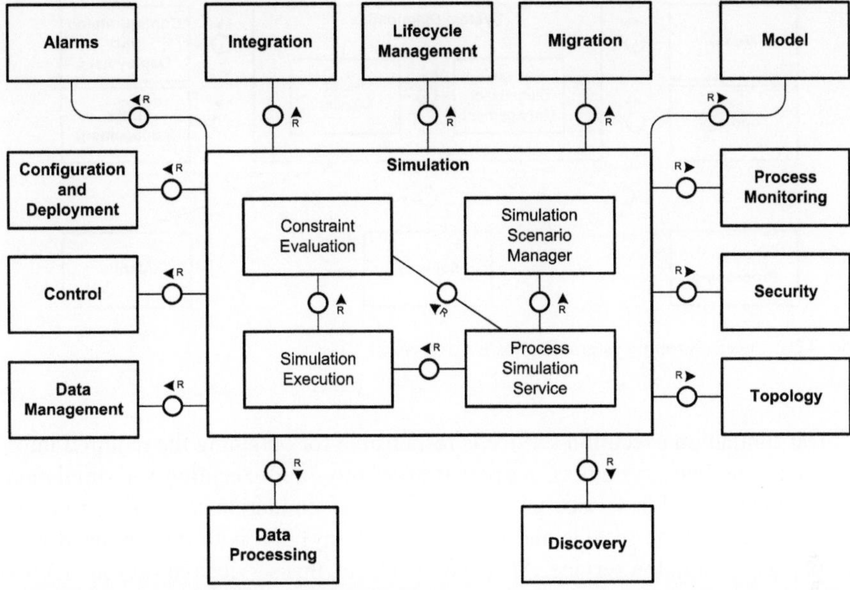

Fig. 3.19 Service group: simulation overview

provided by other services to emulate the performance and behaviour of a system (or a multitude of systems). It consists of four main services: constraint evaluation, simulation scenario manager, simulation execution and process simulation service.

The constraint evaluation service validates a given model with associated constraints and returns possible solutions of the constraint system if those solutions do exist. An example scenario would be the distribution of processes to given topology. The constraint evaluation is able to get as a model the topology with capabilities of the various nodes. In addition, the service needs information about the constraints of process which should be distributed, such as worst-case execution time (WCET), network bandwidth, etc. Based on this information the constraint evaluation can provide a possible distribution of processes on the nodes fulfilling the given constraints.

The process simulation service is in charge of simulating functional and non-functional behaviours of specific processes and validating the feasibility and performance of the simulated processes. This can be used also for operator training. This service interacts with the simulation scenario manager for managing scenario-specific processes to simulate, with the constraint evaluation service to validate processes and with simulation execution service to deploy and execute simulated processes. An example would be where the process simulation tool needs to manage simulated processes. The engineer uses the process simulation tool to create, update or load simulated processes and to validate the processes before simulation execution. The engineer also needs to manage simulated processes under specific simulation scenarios.

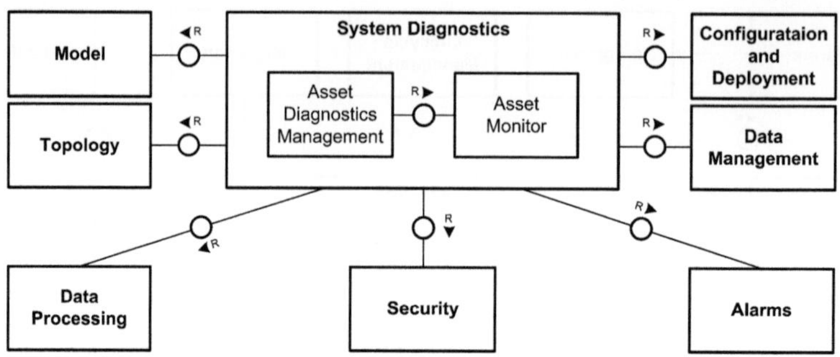

Fig. 3.20 Service group: system diagnostics overview

The simulation execution service is responsible for obtaining the required information to simulate a system(s), or a part of a system(s), and executing said simulation.
Within the simulation service group it requests the simulation constraints from the
constraint evaluation service and the process (if any) that is to be simulated from
the process simulation service. At an external level, this service requires interaction
with others as depicted also in Fig. 3.19. Each of the previously mentioned services
provides the simulation execution service with any information, services, processes,
workflows, models and logs it might need to execute a successful simulation.

The simulation scenario manager is concerned with the configuration and management of different simulation scenarios. It depends, internally, on the process
simulation service and the constraint evaluation service. The simulation scenario
manager can be used to configure and create simulation scenarios. These scenarios
can be obtained to a certain degree by evaluating the constraints of the different systems. By setting theoretical circumstances it is possible to simulate systems under
different situations.

3.3.18 System Diagnostic

The system diagnostic service group (depicted in Fig. 3.20), provides features for
diagnostics of services and devices. Diagnostics can be used to monitor the health
and condition of devices (shop-floor devices, servers, network devices, SCADA
systems and PLC, etc.), and status of services. This service group is used primarily
for maintenance and planning purposes.

The asset diagnostics management service is used for controlling debugging,
logging and testing capabilities. It can also be used to initiate self-test procedures
on a resource. The capabilities of this service include turning on and off debugging
and error logging, manual setting and examining different parameters and rebooting
a device. The asset diagnostics management service can be used for maintenance

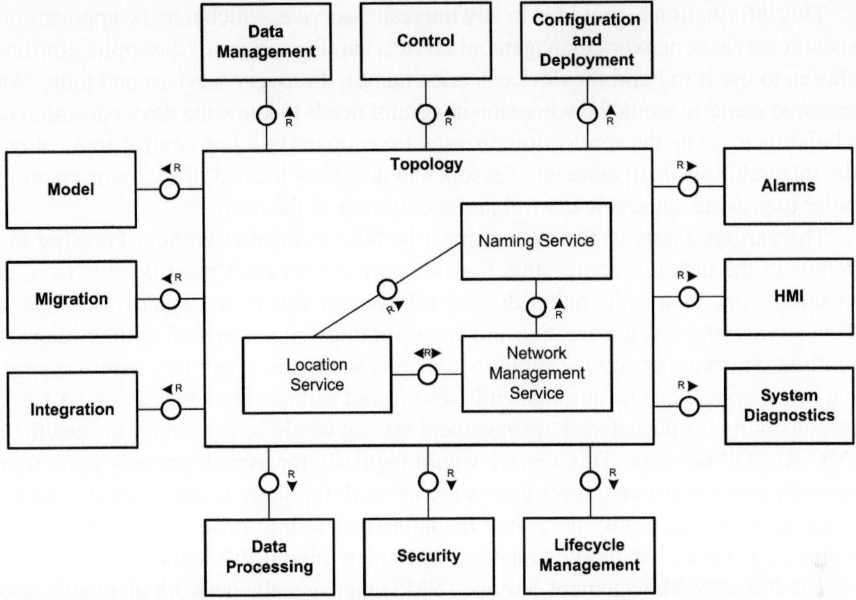

Fig. 3.21 Service group: topology overview

purposes in order to detect faults, initiate self-tests and configure logging of warnings and errors to detect malfunction.

The asset monitor service maintains the current state for each asset. It is also responsible for keeping log of maintenance interventions and planned maintenance schedule. It should be possible to configure the service with specific parameters and characteristics for each asset. These can include operational lifetime, depreciation rate, energy conservation modes, self-testing intervals and safety checks. Based on this information it is possible to perform complex asset management analysis such as risk-based inspections. A possible scenario for the asset monitor service is to provide the foundation for asset life cycle management infrastructure that is capable of optimising the systems operational efficiency in terms of reduced maintenance costs and energy consumption.

3.3.19 Topology

The topology service group (depicted in Fig. 3.21) allows describing and managing the physical and logical structure of the system. It includes Domain Name Service (DNS) functionality, location and context management, network management services, etc.

This information is provided to any interested service, which may be application-specific services, network management service, naming service (if the application has chosen to use it to build the device/service name), discovery services and more. An example scenario would be where the integrator needs to know the device location in a building to set up the application. In order to set up the building control application, the integrator needs to associate sensors and actuators located in the same room in order to provide automatic control and monitoring of the room.

The various assets in the system must be able to interact without knowing the details of the network addressing. For this purpose, an asset must be able to refer to another one by a name rather than by information related to its network address. This service supports the creation and update of these names as well as their usage at runtime. The naming service supports dynamic scenarios where new assets appearing in the system can be automatically discovered and used by other assets. A basic responsibility for the network management service would be to monitor the health of IMC-AESOP network. This service would regularly (or asynchronously upon user request) scan known endpoints across the network topology to assess their connectivity status. Connectivity here may be defined according to several requirements, including network bandwidth/response time and a link/no-link status.

The Network Management Service (NMS) manages the network elements, also called managed devices. Device management includes Faults, Configuration, Accounting, Performance and Security (FCAPS) management. Management tasks include discovering network inventory, monitoring device health and status, providing alerts to conditions that impact system performance, and identification of problems, their source(s) and possible solutions.

The network management service also allows configuring real-time network channels, thus ensuring a proper quality of service for IMC-AESOP real-time services. This service allows configuring the quality of service at router and switch levels through DiffServ and priority queues. IMC-AESOP real-time services will typically require the existence of such real-time channel configuration and management from the underlying networking infrastructure.

The network management service depends on the location service to be able to walk through the entire network topology and on the naming service for simple endpoint identification, independent of any network-addressing scheme. An example scenario would be that of a device failure. A maintenance technician gets an alarm event on his SCADA application. This event has been sent by the network management service upon detection of connectivity loss of a given device. The maintenance technician runs a complete scan of the network to get a detailed health check of the communication layer. This health check should give him enough information to assess the severity of the issue.

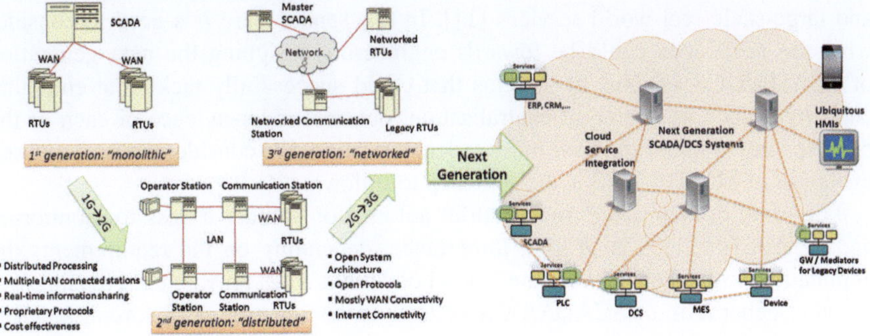

Fig. 3.22 Next-generation SCADA/DCS as a composition in a 'Service Cloud' [11]

3.4 The Next-Generation SCADA/DCS

Service-oriented architectures are considered a promising way towards realising the factory of the future, and we have shown that these can be used to empower infrastructures and their components. The IMC-AESOP architecture and its services already described, offer the possibility of realising the next generation of cyber-physical systems that heavily depend on the cyber part such as cloud-based services. Such an example CPS is the SCADA/DCS systems used today in all industries.

Industrial processes as well as many other critical infrastructures depend on SCADA and DCS systems to perform their complex functionalities. The multitude of functionalities that they need to support as well as the exact roadmap is heavily still researched in an environment where disruptive technologies and concepts are developed rapidly [11]. Having in place an architecture as depicted in Fig. 3.4 has profound implications on the design and deployment of future solutions in the industrial automation domain.

Cyber-physical systems have already undergone significant evolutionary steps in the last decades (shown in Fig. 3.22) and are moving towards an infrastructure that increasingly depends on monitoring the real world, timely evaluation of data acquired and timely applicability of management (control) [11]. The latter is becoming even more difficult to design and manage when massive numbers of networked embedded devices and systems are interacting. As such new approaches are needed that go beyond classical monitoring and are able to deal with massive data and complex reasoning depending on the affected processes as well as enterprise-wide constraints. Such 'capabilities' would by nature require multi-stakeholder involvement and data access that has to go beyond the classical monolithic one-domain and task-specific development approaches.

Currently implemented SCADA/DCS system architectures [2] were designed for more closed and hierarchically controlled industrial environments; however, it is expected that there is potential to enhance their functionality and minimise integration costs by integrating themselves into collaborative approaches with enterprise systems and large-scale real-world services [11]. In this sense, there is a need to consider what the next steps could be towards engineering/designing the next generation of SCADA/DCS systems of systems that could successfully tackle the emerging challenges such as degree of centralisation, optional independence of each of the participating systems and their independent evolution. We consider that cloud-based evolution of SCADA/DCS is the next step to follow.

For some domains, e.g. in industrial automation, timely access to monitoring and control functions is of high importance, depending on the requirements the application poses. For instance the 'Cloud of Things' [12] may be used to empower the next generation of SCADA/DCS systems in conjunction with several services that may be hosted on the devices, in gateways and systems, in the cloud as well as in cross-layer compositions and interactions among them. For many of these, reliability and high-performance interactions are needed, which poses the problem of finding the equilibrium of computation, communication, resource optimisation, openness and user-friendliness in the interactions between the different systems, devices, etc. However for the future we assume that each device or system (generally each 'thing'), can be empowered with Web services either directly (the device is powerful enough to host them locally) or indirectly (the services are provided by a gateway or any other device they are attached to). These services can be accessed directly by applications, systems and other services independent of where they reside empowering a larger collaborative ecosystem of cyber-physical systems such as that envisioned by the IMC-AESOP.

The proposed IMC-AESOP architecture (depicted in Fig. 3.4) could have a significant impact on the way future industrial systems interact and applications are developed. By realising it, a flat information-based infrastructure (as depicted in Fig. 3.1) that coexists with status quo is created. This means that the next-generation SCADA and DCS systems could heavily depend on a set of common services and strike the right balance between functionality co-located on the shop floor and delegated into the cloud [11]. The aim is to have an approach that is fitter for the era where the Internet of Things, infrastructure virtualisation and real-time high-performance solutions are sought. Hence, the next-generation SCADA/DCS systems [11] do not necessarily have to possess a physical nature; this implies that it might reside overwhelmingly on the 'cyber' or 'virtual' world. As such it may comprise multiple real-world devices, on-device and in-network services and service-based collaboration-driven interactions mapped into a 'Service Cloud' (as depicted in Fig. 3.22).

A typical example would be that of asset monitoring with future SCADA systems. In large-scale systems it will be impossible to still do the information acquisition with the traditional methods of pulling the devices, complemented with an event-driven infrastructure. Additionally, sophisticated services would perform analytics on the acquired data, and decision support systems would use their results in real-time to take business relevant decisions. Decision taken will then be enforced enterprise-wide. Such systems will blend from the information flow viewpoint the layers among the different systems and realise the envisioned flat information-driven infrastructure that can be used for mash-up applications and services (as shown in Fig. 3.1).

3.5 Conclusion

Future industrial applications will need to be developed at a rapid pace in order to capture the agility required by modern businesses. Typical industrial software development approaches will need to be adjusted to the new paradigm of distributed complex system software development with main emphasis on collaboration and multi-layer interactions among systems of systems, which is challenging [10]. To do so, some generic common functionality will need to be provided, potentially by a distributed service platform hosting common functionalities, following the service-oriented architecture approach.

Such a collection of services forming a service-based architecture (shown in Fig. 3.4) is presented, prioritised and their potential impact is analysed. Significant effort needs to be invested towards further investigating the interdependencies and needs of all targeted service domains as well as the technologies for realising them. The proposed service architecture attempts to cover the basic needs for monitoring, management, data handling and integration, etc., by taking into consideration the disruptive technologies [11] and concepts that could empower future industrial systems.

Acknowledgments The authors thank the European Commission for their support, and the partners of the EU FP7 project IMC-AESOP (www.imc-aesop.eu) for fruitful discussions.

References

1. acatech (2011) Cyber-physical systems: driving force for innovation in mobility, health, energy and production. Techical report, acatech—National Academy of Science and Engineering. http://www.acatech.de/fileadmin/user_upload/Baumstruktur_nach_Website/Acatech/root/de/ Publikationen/Stellungnahmen/acatech_POSITION_CPS_Englisch_WEB.pdf
2. Barr D (2004) Supervisory control and data acquisition (SCADA) systems. Technical information bulletin 04–1, National Communications System (NCS). http://www.ncs.gov/library/ tech_bulletins/2004/tib_04-1.pdf
3. Colombo AW, Karnouskos S (2009) Towards the factory of the future: a service-oriented cross-layer infrastructure. In: ICT shaping the world: a scientific view. European Telecommunications Standards Institute (ETSI), Wiley, New York, pp 65–81
4. Colombo AW, Karnouskos S, Bangemann T (2013) A system of systems view on collaborative industrial automation. In: IEEE international conference on industrial technology (ICIT 2013), Cape Town, South Africa
5. Delsing J, Eliasson J, Kyusakov R, Colombo AW, Jammes F, Nessaether J, Karnouskos S, Diedrich C (2011) A migration approach towards a SOA-based next generation process control and monitoring. In: 37th annual conference of the IEEE industrial electronics society (IECON 2011), Melbourne, Australia
6. Delsing J, Rosenqvist F, Carlsson O, Colombo AW, Bangemann T (2012) Migration of industrial process control systems into service oriented architecture. In: 38th annual conference of the IEEE industrial electronics society (IECON 2012), Montréal, Canada
7. Drath R, Barth M (2011) Concept for interoperability between independent engineering tools of heterogeneous disciplines. In: IEEE 16th conference on emerging technologies factory automation (ETFA), 2011, pp 1–8. doi:10.1109/ETFA.2011.6058975

8. Jamshidi M (ed) (2008) Systems of systems engineering: principles and applications. CRC Press, Boca Raton
9. Kagermann H, Wahlster W, Helbig J (2013) Recommendations for implementing the strategic initiative INDUSTRIE 4.0. Techical report, acatech—National Academy of Science and Engineering. http://www.acatech.de/fileadmin/user_upload/Baumstruktur_nach_Website/Acatech/root/de/Material_fuer_Sonderseiten/Industrie_4.0/Final_report__Industrie_4.0_accessible.pdf
10. Karnouskos S (2011) Cyber-physical systems in the SmartGrid. In: IEEE 9th international conference on industrial informatics (INDIN), Lisbon, Portugal
11. Karnouskos S, Colombo AW (2011) Architecting the next generation of service-based SCADA/DCS system of systems. In: 37th annual conference of the IEEE industrial electronics society (IECON 2011), Melbourne, Australia
12. Karnouskos S, Somlev V (2013) Performance assessment of integration in the cloud of things via web services. In: IEEE international conference on industrial technology (ICIT 2013), Cape Town, South Africa
13. Karnouskos S, Colombo AW, Jammes F, Delsing J, Bangemann T (2010) Towards an architecture for service-oriented process monitoring and control. In: 36th annual conference of the IEEE industrial electronics society (IECON 2010), Phoenix, AZ
14. Karnouskos S, Savio D, Spiess P, Guinard D, Trifa V, Baecker O (2010) Real world service interaction with enterprise systems in dynamic manufacturing environments. In: Artificial intelligence techniques for networked manufacturing enterprises management. Springer, London
15. Karnouskos S, Vilaseñor V, Handte M, Marrón PJ (2011) Ubiquitous integration of cooperating objects. Int J Next Gener Comput 2(3):2
16. Karnouskos S, Colombo AW, Bangemann T, Manninen K, Camp R, Tilly M, Stluka P, Jammes F, Delsing J, Eliasson J (2012) A SOA-based architecture for empowering future collaborative cloud-based industrial automation. In: 38th annual conference of the IEEE industrial electronics society (IECON 2012), Montréal, Canada
17. Northrop L, Feiler P, Gabriel RP, Goodenough J, Linger R, Longstaff T, Kazman R, Klein M, Schmidt D, Sullivan K, Wallnau K (2006) Ultra-large-scale systems—the software challenge of the future. Technical report, Software Engineering Institute, Carnegie Mellon. http://www.sei.cmu.edu/library/assets/ULS_Book20062.pdf
18. Tranquillini S, Spiess P, Daniel F, Karnouskos S, Casati F, Oertel N, Mottola L, Oppermann FJ, Picco GP, Römer K, Voigt T (2012) Process-based design and integration of wireless sensor network applications. In: 10th international conference on business process management (BPM), Tallinn, Estonia
19. Xu LD (2011) Enterprise systems: state-of-the-art and future trends. IEEE Trans Industr Inf 7(4):630–640. doi:10.1109/TII.2011.2167156

Chapter 4
Promising Technologies for SOA-Based Industrial Automation Systems

François Jammes, Stamatis Karnouskos, Bernard Bony, Philippe Nappey, Armando W. Colombo, Jerker Delsing, Jens Eliasson, Rumen Kyusakov, Petr Stluka, Marcel Tilly and Thomas Bangemann

Abstract In the last years service-oriented architectures have been extensively used to enable seamless interaction and integration among the various heterogeneous systems and devices found in modern factories. The emerging Industrial Automation Systems are increasingly utilising them. In the cloud-based vision of IMC-AESOP such technologies take an even more key role as they empower the backbone of the new concepts and approaches under development. Here we report about the investigations and assessments performed to find answers to some of the major questions that arise as key when technologies have to be selected and used in an industrial context utilizing Service-Oriented Architecture (SOA)-based distributed large-scale process monitoring and control system. Aspects of integration, real-timeness, distributeness, event-based interaction, service-enablement, etc., are approached from different angles and some of the promising technologies are analysed and assessed.

F. Jammes (✉) · B. Bony · P. Nappey
Schneider Electric, Grenoble, France
e-mail: francois2.jammes@schneider-electric.com

B. Bony
e-mail: bernard.bony@schneider-electric.com

P. Nappey
e-mail: philippe.nappey@schneider-electric.com

S. Karnouskos
SAP, Karlsruhe, Germany
e-mail: stamatis.karnouskos@sap.com

A. W. Colombo
Schneider Electric, Marktheidenfeld, Germany
e-mail: armando.colombo@schneider-electric.com

A. W. Colombo
University of Applied Sciences Emden/Leer, Emden, Germany
e-mail: awcolombo@technik-emden.de

A. W. Colombo et al. (eds.), *Industrial Cloud-Based Cyber-Physical Systems*,
DOI: 10.1007/978-3-319-05624-1_4, © Springer International Publishing Switzerland 2014

4.1 Introduction

Current industrial process control and monitoring applications are facing many challenges as the complexity of systems increases and the systems evolve from synchronous to asynchronous. When hundreds of thousands of devices and service-oriented systems are asynchronously interconnected and share and exchange data and information, i.e., services, for monitoring, controlling, and managing the processes, key challenges such as interoperability, real-time performance constraints, among others, arise and need to be addressed.

The SOA-based approach proposed by the European R&D projects SOCRADES and subsequently IMC-AESOP [12], addresses some of these challenges. The vision pursued is shown in Fig. 4.1, according to which the industrial process environment is mapped into a 'Service Cloud', i.e. devices and applications distributed across the different layers of the enterprise expose their characteristics and functionalities as 'services'. Additionally, these devices and systems are able to access and use those 'services' located in the cloud [10, 11, 13].

The outcomes of the first set of industry technology investigations and pilot applications, carried out according to the IMC-AESOP project vision [12, 13], reveals four major challenges that may need to be addressed:

I Real-time SOA: Determine the real-time limits of bringing SOA inside the high performance control loops of process monitoring and control (e.g. Is it possible to provide service-oriented solutions targeting the one millisecond performance range?)

II Large-scale distributed process control and monitoring system: Is it feasible to dynamically design, deploy, configure, manage and maintain an open plant/enterprise wide system, with thousands of devices and systems operating

J. Delsing · J. Eliasson · R. Kyusakov
Luleå University of Technology, Luleå, Sweden
e-mail: jerker.delsing@ltu.se

J. Eliasson
e-mail: jens.eliasson@ltu.se

R. Kyusakov
e-mail: rumen.kyusakov@ltu.se

P. Stluka
Honeywell, Prague, Czech Republic
e-mail: petr.stluka@honeywell.com

M. Tilly
Microsoft, Unterschleißheim, Germany
e-mail: marcel.tilly@microsoft.com

T. Bangemann
ifak, Magdeburg, Germany
e-mail: thomas.bangemann@ifak.eu

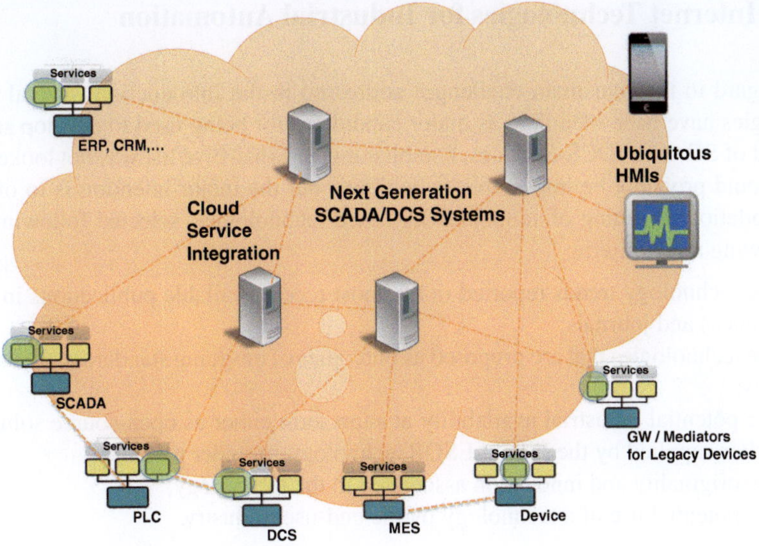

Fig. 4.1 IMC-AESOP Approach: a distributed dynamically collaborative service-oriented SCADA/DCS system

under process real-time constraints and still comply to ISA-95 (http://www.isa-95.com) and PERA (http://www.pera-net) architectures?

III Process monitoring and control systems operating in an asynchronous mode, e.g. Distributed event-based systems: Which are the technological consequences and limits of these asynchronous SCADA/DCS platforms when compared to traditional implemented periodic systems? Is it possible to integrate asynchronous and synchronous systems, e.g. for legacy system integration?

IV Service specification: Which methodology and tools are the most suitable to identify and specify the semantics for interoperable (standard/common/specific) Web services-based monitoring and control (from business process to devices)?

In this work we present the results of the investigations and assessments performed [8] to find some of the answers to those four guiding questions when technologies have to be selected and used in a Service-Oriented Architecture (SOA)-based distributed large-scale process control and monitoring system. First, we present a description and assessment of the most suitable technologies for addressing the four challenges described above in the area of industrial automation. Subsequently, the results of the assessment synthesising the technologies that are being used to implement the IMC-AESOP approach are shown together with highlights and some outlook for the future.

4.2 Internet Technologies for Industrial Automation

In regard to the four main challenges addressed in the Introduction, several technologies have been identified as major candidates for being used to develop such a cloud of SCADA-/DCS-Services. Establishing an exhaustive list was not looked for (it would probably be impossible to achieve), but the major intention is to offer a compilation/screening of suitable SOA-based technologies, selected following the following main criteria:

- The technology trends reported in the most recent available publications in conferences and journals;
- The technologies that are proposed as outcomes of on-going standardisation activities;
- The potential industrial availability at short term either as open-source solutions and/or supplied by the IMC-AESOP technology-provider partners;
- The originality and innovation associated to the technology;
- The potential use of a technology by the end-user industry.

Some key technologies identified include: DPWS, EXI, CoAP, REST, OPC-UA, Distributed Service Bus, Complex Event Processing (CEP), Semantic Technologies.

4.2.1 DPWS and EXI

DPWS is recognised as a good SOA device level protocol profile. Among all Web service protocols, it selects the most appropriate ones, such as WS-Discovery and WS-Eventing above SOAP, for implementation in constrained embedded devices. It provides capabilities such as interoperability, plug and play, and integration capability. Several projects such as SOCRADES (http://www.socrades.eu) and SIRENA http://www.sirena-itea.org have demonstrated its capabilities. However, it does not provide real-time performance in the millisecond range [6] alone. However, if we couple DPWS with EXI (www.w3.org/XML/EXI), this performance target is achievable.

When looking at the real-time challenge, the performance that is evaluated and measured is defined as the time to send an event from one device application to another remote device application on a local network. This is done taking into account the time periods required to go through emitter and receiver stacks and to go through the local network, in a one-way asynchronous event transmission.

In the example shown in Fig. 4.2, two remote devices are connected by a physical network (e.g. Ethernet). The first device detects a data change on one of its physical inputs, and sends this information through the network to the second device, which then generates a corresponding physical output. Both devices use DPWS [4] to exchange the information, which provides all the customer values of DPWS (interoperability, plug and play, integration capability). They integrate inside the DPWS stack the EXI encoder or decoder capability to add real-time performance to the standard DPWS values.

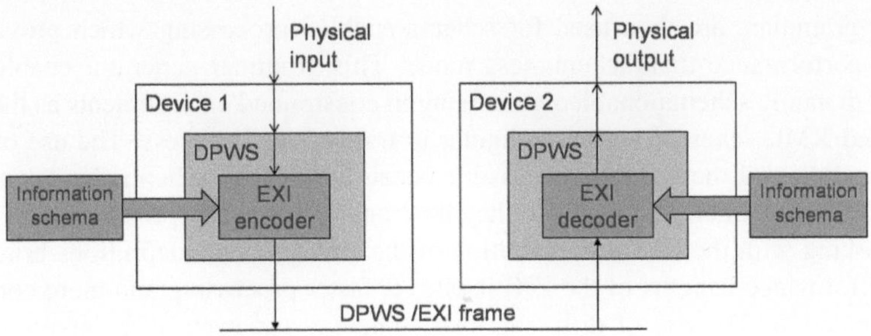

Fig. 4.2 DPWS / EXI integration

After this exchange, the first device, when receiving an input change, will translate this physical event into a DPWS/EXI network event, using the combined capabilities of the DPWS stack and the EXI encoder, which was programmed or configured according to the information schema. The second device, when receiving the network event, will decode the frame and transform it into an output change.

4.2.2 EXIP: EXI Project

The implementation of XML/EXI technology provides a generic framework for describing, implementing and maintaining complex systems and interactions. However, the usage of XML, even when used with a binary compressed representation, can result in a too high overhead for deeply constrained devices. Furthermore, the application of complex schemas and WSDL descriptions can make versioning difficult since the XML/EXI parsers might require updated grammars for optimal performance.

In some cases, the use of a more simple data representation such as JSON and SenML might be sufficient, especially for very low-cost sensors and actuators. However, the implementation of different data representation techniques between the resource constrained devices and more capable systems requires service gateways that convert these data formats. Using service gateways and mediators introduces complexity in the provisioning and maintenance of the systems. In such scenario, it is beneficial to use EXI all the way down to the sensor and actuator devices.

Although the EXI format is designed for high compression and fast processing, its deployment on deeply constrained devices such as wireless sensor nodes is challenging due to RAM and programming memory requirements. The EXIP open-source project [14] is providing efficient EXI processing for such embedded devices. The EXIP prototype implementation is specially designed to handle typed data and small EXI messages efficiently as this is often required in process monitoring and control applications for sensor data acquisition.

The EXIP project also includes a novel EXI grammar generator that efficiently converts an EXI encoded XML schema document into EXI grammar definitions.

These grammars are then used for schema-enabled processing which provides a better performance than schema-less mode. This grammar generator enables the use of dynamic schema-enabled processing in constrained environments as the EXI encoded XML schemas are much lighter to transmit and process. The use of EXI representation of the schemas is possible because the XML schema documents are plain XML documents and as such they have analogous EXI representation.

Working with the EXI representation of the XML schema definitions brings all the performance benefits of the EXI itself, i.e. faster processing and more compact representation. The use of different XML schemas and even different version of these schemas at runtime is challenging. For that reason, an important future work investigation is the support for XML schema evolutions in the SOA implementations.

Another important aspect is the definition of EXI profile for implementation in industrial environments that will guarantee interoperability and optimal performance of the EXI processing. This profile must specify what options should be used in the EXI headers and how the schema information is communicated between the devices and systems.

The main results of the performed evaluation of EXI show that:

- The use of EXI provides significant reduction in the exchanged message sizes. Compression ratios up to 20-fold may be obtained for some types of messages. Although the experiment has been performed on a high-speed wired Ethernet network, it is also expected that low-bandwidth networks such as those found in wireless sensor networks would also strongly benefit from the use of EXI.
- Performance improvements are less significant: only an improvement by a factor of 2 has been measured. This is due in part to the inherent complexity of EXI, which is computation-intensive, but also to the overhead of the underlying message exchange protocols (HTTP and SOAP in the experiment). Further experiments using more efficient protocols, such as a simple TCP protocol or the new CoAP protocol, could demonstrate that EXI is also relevant for high-performance applications.

4.2.3 CoAP

In the era of lightweight integration, especially of resource-constraint devices with web technologies, a new application protocol is proposed within the Internet Engineering Task Force (IETF), i.e. the Constrained Application Protocol (CoAP) [2, 21]. CoAP provides a method/response interaction model between application endpoints, supports built-in resource discovery and includes key web concepts such as URIs and content-types. CoAP also easily translates into HTTP for seamless integration with the Web, while meeting specialised requirements such as event-based communication, multicast support, very low overhead and simplicity for constrained environments.

As depicted in Fig. 4.3, CoAP relies on UDP instead of TCP that is used by default for HTTP integration. UDP provides advantages for low overhead and multicast

Fig. 4.3 CoAP lightweight integration versus heavy HTTP integration

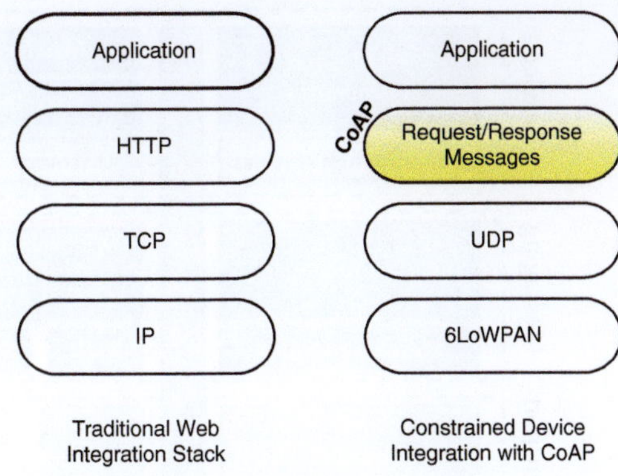

support. CoAP is REST centric (supports GET, POST, PUT, DELETE), and although it can be used to compress HTTP interfaces it offers additional functionalities such as built-in discovery, eventing, multicast support and asynchronous message exchanges. From the security point of view several approaches are supported ranging from no-security up to certificate-based using DTLS. IANA has assigned the port number 5683 and the service name 'CoAP'.

Within the IMC-AESOP project, CoAP is mainly considered to get access to extremely resource constraint devices, e.g. a temperature sensor, a wireless sensor node, etc. Moreover, the devices may also be mobile and rely on a battery for their operation. These distributed devices would probably be used for monitoring and management, while their integration may enhance the quality of information reaching SCADA/DCS systems.

4.2.4 OPC-UA

One of the challenges in process industries is the interoperability between systems and devices coming from numerous vendors. This has been addressed by using open standards, enabling devices from different vendors to understand each other. One of the widely accepted standards is OPC (OLE for process control). However, after many years of its use, some limitations of this standard have become evident. This was the reason the OPC Foundation started to work on the new standard—OPC Unified Architecture (OPC-UA) [16].

OPC-UA main improvements over the classic OPC include the following:

- Unified access to existing OPC data models (OPC DA, OPC HDA, OPC A/E, etc.);
- Multi-platform implementations;
- Communication and security (OPC has been based on COM/DCOM);
- Data modelling.

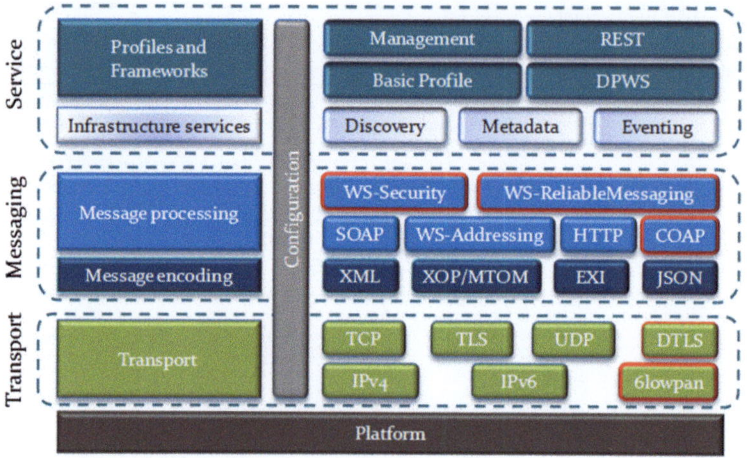

Fig. 4.4 ESF modular architecture

While the communication, security and interoperability features make OPC-UA a great candidate to be used in SOA-based applications, it is its data modelling capabilities that enable to build a service-oriented process control system [24]. OPC-UA provides means to access not only the data from the process systems, but also semantic information related to the data, like models of the devices that are providing this data. Such models are built by defining nodes (described by attributes) and relations between the nodes (Fig. 4.4).

An information model contains definition of types, from simple to complex, and also instances of such types. The information models are organised and exposed by address spaces. In an existing implementation, multiple information models can be defined; for each level on the process there can be a different model of the process entities, although these models can share information and are usually synchronised. With growing penetration of OPC-UA into the processes and its features that have been designed with SOA in mind, it is clear that OPC-UA will become a solid part of service-oriented distributed control systems.

4.3 Technology Combinations and Advanced Concepts

Apart from the basic technologies, we take a closer look at efforts for their convergence and provision of more advanced functionalities for future industrial automation systems.

4.3.1 The Embedded Service Framework

The Embedded Service Framework (ESF) is a redesigned, rewritten and extended version of the DPWSCore stack, which is available at http://forge.SOA4d.org. The goals of this new version are:

- To bring the power of recent Web-oriented technologies to embedded devices utilising service-oriented and REST architectures.
- To hide complexity from developers, through code generation and high-level APIs.
- To support a large range of applications, from basic Web applications to complex Web service applications, featuring mechanisms such as network discovery and event publishing.
- To support a wide range of platforms, from mono-threaded (or OS-less), deeply embedded devices (e.g. wireless sensors) to complex multi-threaded applications running on large devices, workstations or enterprise servers.

The main features of the ESF include:

- Support of standard IPv4 and IPv6-based transport protocols: TCP, UDP, TLS. Support for additional protocols such as 6LoWPAN and DTLS is also planned.
- Support of standard message encodings: Besides XML, which is used in several standard messaging protocols, ESF also supports XOP/MTOM, used to transport binary data in SOAP messages, EXI, a standard binary format for XML well-adapted to low-bandwidth networks, and JSON, a popular format used in particular in browser-based applications.
- Support of several messaging protocols, including HTTP, SOAP (directly over TCP or UDP or combined with HTTP) and SOAP extensions such as WS-Addressing. Planned additions include WS-Security and WS-ReliableMessaging, or CoAP, a draft IETF standard designed for REST applications over 6LoWPAN, but also usable over standard IP networks.
- A set of infrastructure services for network discovery of devices and services, metadata exchange, event publication and subscription or resource management. These services allow the implementation of standard profiles such as Basic Profile (1.1 and 2.0) and Devices Profile for Web Services (DPWS), or of resource management frameworks such as WS-Management or ad hoc solutions based on the REST paradigm.
- A configuration mechanism allowing developers to select the appropriate components from the above list for their applications.

The set of technologies, profiles and frameworks shown in the diagram is not exhaustive: the ESF is extensible and may be used to implement other popular profiles, such as UPnP or ONVIF. Application development on top of the ESF combines access to the runtime library through the ESF API and use of generated code (as shown in). Both client-side and server-side applications may be developed, both sides being often combined in devices capable of peer-to-peer interactions (Fig. 4.5).

On the server side, two paradigms are supported:

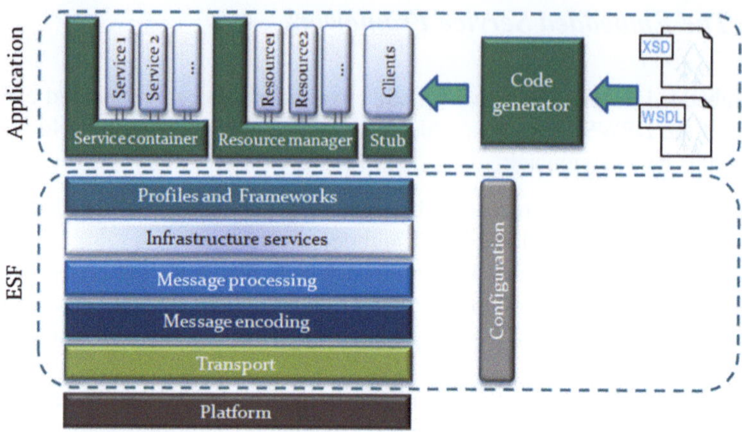

Fig. 4.5 Application development with ESF

- The service-oriented paradigm: Based on the abstract definition of a service interface, through a WSDL document, the code generator produces a service skeleton
  ready to be plugged in the ESF service container. The role of the developer is to
  provide the implementation of the service operations and to configure the ESF
  runtime with the required protocols. ESF makes it easy to publish simultaneously
  the same services using different protocols, in order to extend the reach of those
  services to a wide range of clients.
- The resource-oriented paradigm: The ESF provides a resource manager that allows
  developers to register resource implementations, and provides remote access to
  those resources through REST or Web services protocols.

On the client side, code generation is used to produce service stubs, which can be
used by the application to invoke remote service operations or resource access. The
configurability of the ESF protocols allows clients to access a wide range of devices.

In order to use EXI while guaranteeing the same level of interoperability as XML,
several approaches can be considered:

- *Use of a globally shared configuration*: In stable and managed environments, it
  is possible to deploy the same set of XML schemas in the server and all clients.
  This single EXI configuration is then used to encode and decode all exchanged
  messages. This approach has the drawback of being slightly less efficient, as the
  set of global elements used to encode a given message is larger than needed. On
  the other hand, it simplifies the configuration of the server and the clients.
- *Use of out-of-band information to select the appropriate EXI configuration*: On
  the server side, it is possible to use external information, such as the HTTP request
  URL (e.g. when using SOAP-over-HTTP) or the network listener port (e.g. when
  using SOAP-over-UDP) to select the appropriate EXI configuration. A typical
  deployment configuration would use different HTTP endpoints for different services and SOAP bindings, and associate to each endpoint the minimal set of XML
  schemas needed to parse the incoming EXI messages.

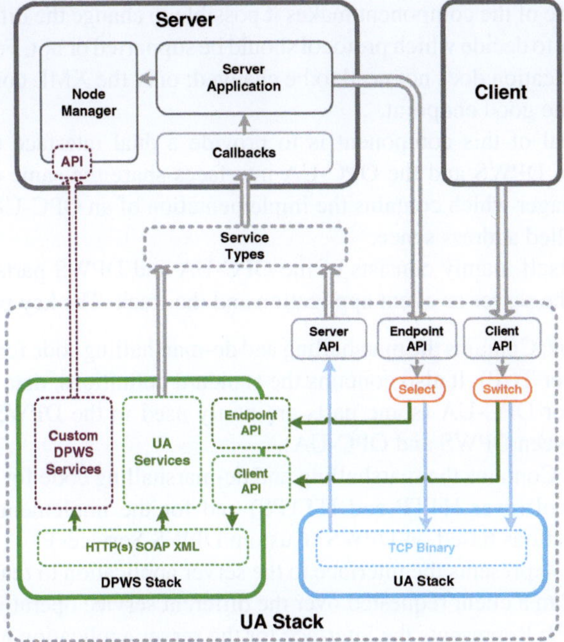

Fig. 4.6 Fusion of OPC-UA and DPWS architecture

- *Use of EXI options*: EXI provides an in-line header mechanism which allows additional data to be communicated to the EXI processor before it starts decoding, among which is a Schema ID. By defining a system-wide naming mechanism for EXI configurations, it is possible to use this solution to dynamically select the appropriate configuration to be used for a given message. This approach has the drawback of slightly increasing the message size, as the Schema ID is systematically embedded at the beginning of each message.

4.3.2 Fusion of DPWS and OPC-UA

As OPC-UA and DPWS have a large set of similarities, it is possible to build a common stack compliant with both standards where the two technologies can benefit from each other. A component implementing the convergence between OPC-UA and DPWS for embedded devices has been prototyped as described in Fig. 4.6.

This component includes in particular:

- The OPC-UA stack developed in ANSI C language by the OPC Foundation and which supports the UA binary profiles defined by the OPC-UA specification.
- The DPWS stack for implementing the Web services profile of OPC-UA.

The architecture of the component makes it possible to change the different libraries of the UA stack to decide which protocol should be supported or not. For this purpose, the server application does not need to be changed; only the XML configuration file must include the good endpoint.

Another goal of this component is to provide a dual interface (i.e. DPWS + OPC-UA). The DPWS and the OPC-UA interfaces share the same data, managed by a node manager which contains the implementation of an OPC-UA enabled data model, also called address space.

The stack itself mainly consists of the OPC-UA and DPWS parts and a unified API between the server or client application and the stack. The key parts are:

- *OPC-UA part*: Contains the marshalling and de-marshalling code for the UA binary protocol (over TCP). It also contains the standard definitions, data structures and data types for OPC-UA (some parts are partly used in the DPWS part to get a binding between DPWS and OPC-UA).
- *DPWS part*: Contains the marshalling and de-marshalling code for the UA SOAP XML protocol (over HTTP and HTTPS) and for the implementation of other service operations based on DPWS (Custom DPWS Services).
- *Server API*: Represents the interface to the server application to react to incoming messages from a client requested over the different service operations.
- *Endpoint API*: Represents the interface for the server application to manage endpoints (Create, Open, Close, Delete,…).
- *Endpoint API for DPWS part*: Represents the interface for the DPWS stack to manage endpoints (internal API). The design is related to Endpoint API of the final stack which can be called from outside.
- *Client API*: Represents the interface for the client application to use service operations for the communication with a server.
- *Client API for DPWS part*: Represents the interface for the DPWS stack for using the supported client operations (internal API).
- *Service Types*: Responsible to call the correct callback function in the server application concerning the called service operation from a client. More information about the service types is given in the following chapter.

The following features implemented and tested, show that the DPWS stack can be used for implementing an HTTP/HTTPS profile for an OPC-UA stack and that the resulting component can expose both an OPC-UA and a DPWS interface:

- Communication over HTTP SOAP XML profile is working.
- Communication over OPC TCP Binary profile is working.
- Communication over HTTPS is working.
- Server can be used to deploy a predefined XML data model description for a device.
- Custom Web services can be discovered and called in conformance with the DPWS specification.

The DPWS/OPC-UA prototype has demonstrated promising benefits for systems with a large number of devices, in particular when the data exposed by the devices are heterogeneous. In the following we consider a system including a client application and a set of devices or subsystems where all communicating entities are implementing a converged stack with DPWS and OPC-UA: DPWS brings the capability for the client application to discover dynamically a large amount of devices. We have tested that at least 1,000 devices can be discovered at the same time.

The discovered devices have then to expose their data to the client application. Even if the semantics of the data exposed by all the devices are heterogeneous, the data can be individually mapped on the generic meta-model of OPC-UA. This can be done through proprietary mappings or preferably by mappings already specified and validated by the OPC Foundation (OPC-UA companion standards). For the OPC-UA enabled client application, the result is that it can understand the data exposed by all the devices. This client application may be either a completely generic OPC-UA application, in which case it will understand the data with a limited semantic level, or the client application may be more aware of the domain semantic (either proprietary or defined in OPC-UA companion standards).

4.3.3 Distributed Service Bus

Web service-based technologies investigated so far at device level (DPWS, OPC-UA, etc.) rely mainly on point-to-point communication models, which do not favour the system scalability. The 'Service Bus' approach aims at decoupling service consumers from service producers in the industrial process control system. Large-scale distributed systems can benefit from a service bus-type middleware architecture as the bus acts as a broker between the numerous service consumers/providers, avoiding a potentially huge number of point-to-point connections.

The service bus middleware (depicted in Fig. 4.7) is based on a distributed architecture to share information between all middleware instances. In other terms, devices and systems handled by an instance of the service bus are exposed through a normalised data model and this information is shared with the other instances.

Legacy systems can also benefit from service bus architecture as the bus acts as a gateway between legacy systems and IMC-AESOP SOA systems. This service bus is therefore the natural place for adding a semantic layer on top of legacy services. Thus, the bus provides an abstraction of technical devices and services into business-oriented/domain-specific service descriptions.

Figure 4.8 gives a functional view of the distributed service bus and illustrates how it hosts some of the services identified in the IMC-AESOP architecture study, for instance:

- Gateway functionality through a variety of connectors;
- Registry as a central repository for IMC-AESOP services;
- Code/configuration/model repository (not implemented yet);
- Event broker for true loose-coupling between event producers and consumers;
- Security services;
- DNS service;

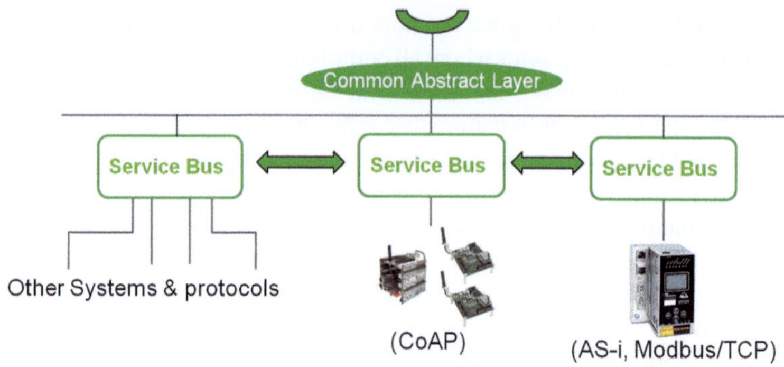

Fig. 4.7 Using the service bus as common abstraction layer

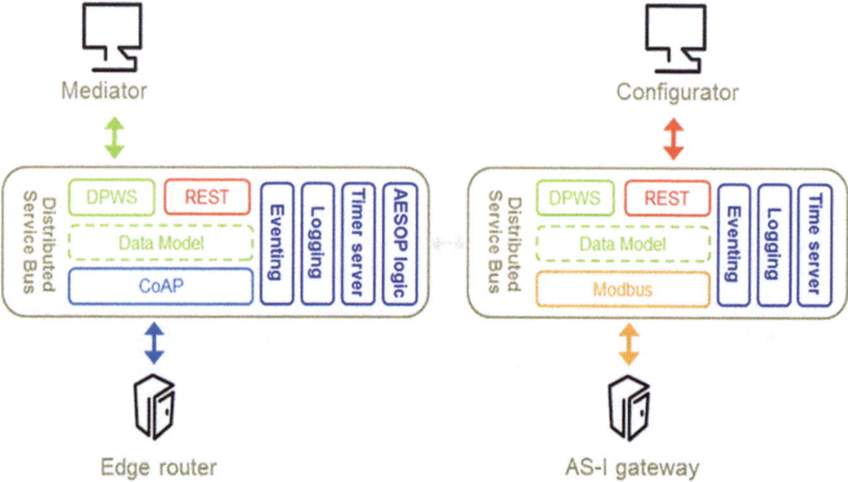

Fig. 4.8 Distributed service bus architecture

- Historian/logger;
- Time service for time synchronisation between IMC-AESOP services;
- Native interface (Web services) to higher level information systems (MES/ERP...).

The modularity of the service bus allows adding protocol connectors and application modules to manage various devices and services. Such management operations are applied through a common abstract layer. The distributed architecture of the service bus allows a management operation to be routed to the adequate service bus instance handling the targeted device or service. Therefore, the distributed architecture of the service bus and the common interface through the abstract layer both enable the management of large-scale systems.

The service bus implementation is currently available in C and Java languages. The C brick can be embedded in devices with constrained resources; it requires around 200

KB of Flash memory and 50 kb of RAM with all connectors/modules included. The Java brick requires obviously much more resources and will run on more powerful devices able to run a Virtual Machine. This is the gateways and controllers that can be found in typical process control systems. The Java implementation relies on the OSGi Framework 'Felix' (http://www.felix.apache.org). OSGi is an SOA-based modular framework which implements a dynamic component model. It also provides dynamic life cycle management for its modules which can be started, stopped, updated without an application reboot. This capability is particularly interesting for high-end service bus instances, where new modules and connectors can be deployed at runtime.

The service bus provides several set of management operations which can be applied to devices and services. Device management capabilities include adding/removing/discovering devices, getting/setting configuration and status. Typical service management includes start/stop/reset of services, getting/setting configurations and status. The device and service management operations are accessible through both SOAP and REST interfaces.

The distribution among service buses is also handled through an internal DPWS/SOAP interface. This DPWS interface handles the mutual discovery between service bus instances thanks to WS-Discovery.

Time synchronisation relies on the IEEE 1588 PTP (Precision Time Protocol) protocol running on all service bus instances. Time synchronisation is a strong requirement for events timestamping and correlation. Events logging is implemented based on the standard syslog protocol, which allows to aggregate events from all service bus instances on a central repository. This is particularly useful for correlating system events, for root cause analysis for instance. This capability has been used in particular in use case 1, providing useful insights into system behaviour.

Cyber-security was not at the heart of the IMC-AESOP project so only minimal support was provided through the service bus component. This includes user authentication through HTTP basic authentication and a simplified Role-Based Access Control (RBAC) applied to each service call. Practically, only admin users were able to invoke services from the service bus.

The service bus can handle large-scale distribution by relying on its connector's distribution. Each connector exposes devices and services in a common abstract way. Moreover, information from the abstract layer is actually exchanged between all instances of the service bus. Such distribution allows any application to interact with the real devices/services transparently through a common interface which is provided by any service bus instance.

4.3.4 Complex Event Processing

Throughout the last years, CEP [15] has gained considerable importance as a means to extract information from distributed event-based (or message-based) systems. It became popular in the domain of business process management but is now applied in the industrial monitoring and control domains. It is a technology to derive higher

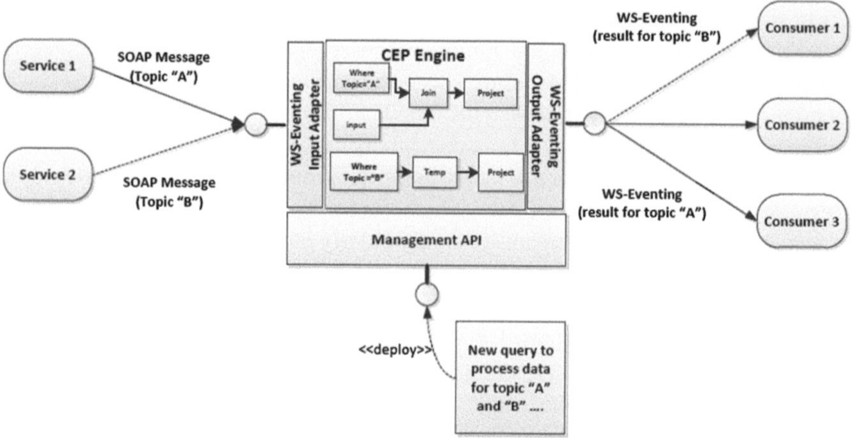

Fig. 4.9 Complex event processing mechanism in an SOA-Infrastructure

level information out of low-level events. CEP relies on a set of tools and techniques for analysing and handling events with very low latency. The feature set for CEP spans from event extraction, sampling, filtering correlation and aggregation to event enrichment, content-based routing, event compositions (and are not limited to these).

Originally, CEP systems were created at enterprise systems; therefore, most available systems provide tools to define queries and to manage and administrate the system. Some of them provide concepts for scalability and resilience. By contrast, nowadays, we can observe a trend to move CEP closer to the place where the data are born to enable early filtering, aggregation and resampling capabilities. In that way it will become possible to write or define a query and distribute it seamless cross a distributed set-up in a way to reduce network traffic and save bandwidth.

Normally, complex events are created by abstracting from low-level events. The processing of events is expressed within a specific language in terms of rules. Unfortunately, the set of features and the way to express the rules differ from platform to platform. CEP engines are able to process events up to 100,000 events/s. This clearly depends on the complexity of the rules. Normally the limitation is set by the connection to the external environment, such as extraction of events from input sources or the limitation by the bandwidth of the network.

So far, there is no unified way to express rules (or queries) over streams of events. Thus, it makes sense to wrap a CEP engine (Fig. 4.9) within a service with well-defined endpoints. The endpoints are technology agnostic and define the operations and data to be processed while the CEP service itself is responsible for transforming the data/messages to its internal event format. On the output side consumers can subscribe via WS-Eventing so that notifications can be sent via SOAP messages as well (see Fig. 4.8). This approach enables the integration [7] with specifications like Device Profile for Web services (DWPS) and OPC Unified Architecture (OPC-UA), which are the most suitable solutions for implementing an SOA since both specifications include eventing mechanisms.

Two kinds of CEP are expected to be provided in future industrial systems:

- *CEP as a service*: When we say service, it means that this can either be realised as a service running locally on a server or the same concept still holds for a service running in the cloud and on top of cloud technologies.
- *Embedded CEP*: This is a concept of a lightweight CEP using Concurrent Reactive Objects (CRO) model guaranteeing execution of CEP queries in an efficient and predictable manner on resource constrained platforms and offering a low-overhead real-time scheduling.

By enabling event processing mechanisms IMC-AESOP also considers the convergence of scan-based and event-based mechanisms. This is achieved by supporting pull- or push- models [22]. The services can either send events (active) to the CEP service or there is a mediator which pulls data from services (passive) and sends this data. From the CEP service this looks like an active data service provider. On the output side results are pushed to registered consumers.

4.3.5 Semantic-Driven Interaction

Enabling interoperability of the service specifications and data models is a key technological challenge that SOA systems aim to resolve. The full interoperability requires that the syntax and semantic service descriptions are well-defined, unambiguous and enable dynamic discovery and composition. Thus far, most if not all SOA installations enable pure syntax interoperability with little or no support for standard-based semantic descriptions. The use of structured data formats only partially resolves the problem by supplementing the exchanged data with meta-information in the form of tags and attributes in the case of XML/EXI for example. The tag names are ambiguous and usually insufficient to describe the service functionality in full.

Applying application level data model standards is often a remedy as the syntax to semantics mapping is predefined. Example of such standard is Smart Energy Profile 2 that clearly states the physical meaning of the tag names and structures defined for the service messages in the domain of energy management. One problem when complying with such standards is that they are almost always domain specific which requires mapping of the semantic descriptions from one standard to all others in use.

Another approach is to define generic semantic data model that is applicable to a wide range of use cases. The initial investigation highlighted the Sensor Model Language (SensorML) [20] as a promising specification for generic semantic description of sensory data. However, the complexity and size of SensorML specification limit its use to more capable devices. Small-scale experiments with a number of sample SensorML messages showed that even EXI representation will not be sufficiently small to fit battery powered wireless sensor nodes that have low-power, low-bandwidth radios.

Another possible specification for sensor data is the Sensor Markup Language (SenML) [9]. It has a simple design that is consistent with RESTful architecture and

Table 4.1 Technologies and challenges

Technologies	Real-time	Management of large scale	Event-driven	Semantics
DPWS		X	X	
OPC-UA		X		X
CoAP		X	X	
EXI	X		X	
Service bus		X		
CEP			X	

is targeted at resource-constrained devices. An initial evaluation of SenML revealed that it meets the requirements for hardware utilisation but there are areas that are too simplified and insufficient to describe the data in the details required by the applications. An example of such limitation is the precision of the time stamping of the sensor data—SenML allows for up to seconds resolution that is not enough for many industrial use cases. To overcome this limitation, we had to use a custom generic data representation that reuses many of the design choices in SenML.

4.4 Discussion

DPWS, coming from the IT world, is the most applicable set of Web services protocols, to be used at the device level. Combined with EXI, it provides real capabilities in the range of milliseconds, following the technology assessment made by the project. OPC-UA, coming from the industrial world, is also a set of Web service protocols, compatible with DPWS, and providing a data model enlarging the semantic capabilities of the solution. CoAP can be used for wireless sensor networks. It can also be combined with EXI. This is still work-in-progress with major impact in the future once the technology matures.

The service bus and the CEP solution are technologies providing large scale and migration capabilities, combining and processing information coming through DPWS, OPC-UA or legacy protocols, in order to manage large-scale event-based systems. A suitable combination of the six technologies described above is able to provide solutions meeting the four critical questions and challenges expressed in the Introduction.

After some initial assessment and taking into consideration the operational context of IMC-AESOP, we have come up with a synthesis of the most promising technologies (depicted in Table 4.1), which are being used to implement the IMC-AESOP prototypes:

A closer look at some of the Web services-based technologies and their performance [11] reveals several aspects. Although DPWS is already available and supported by several devices in multiple domains, we can clearly see that in its standard form it has a significant impact on computational and communication resources.

Hence devices that may consider this stack, should be usually devices that are on the upper scale with respect to their resource availability. REST and CoAP are designed for much more lean environments and as we see these are a much better fit for resource constrained devices, e.g. in comparison to DPWS [18]. Additional combinations of DPWS with compression techniques however could remove this barrier [17].

REST and CoAP approaches are more lightweight (from CPU and memory utilisation) and more user-friendly implementation-wise, and therefore could empower even simple sensors to take part in the 'Cloud of Things' [11]. On the cloud side, since we do not have significant resource problems, any of the stacks can be used, but maybe for scalability reasons the lightweight REST might also be preferred, unless some specific functionality is needed, e.g. WS-Discovery from the DPWS to dynamically discover embedded devices and their services. Further customisations may enable hybrid approaches such as SOAP over CoAP [19]. Additionally, ongoing work, e.g. in EXI [3] may also enable better performance when combined with the XML-based approaches.

OPC-UA is not a real-time protocol, but is designed rather to gather information about the transferred data with the occurrence time stamp and distribute that information on demand [4]. OPC-UA services are designed for bulk operations to avoid roundtrips, something that increases the complexity of the services but greatly improves the performance [16]. Nevertheless, the balance between functionalities and performance needs to be per scenario investigated, especially due to the multiple other aspects the OPC-UA brings with it as already analysed.

Although the initial tests [11] are not conclusive and offer only a notion of performance, there are several other issues that need to be investigated and which may be of critical importance, depending on the application domain targeted. Security is an issue, and the impact has not been investigated here as we considered only *HTTP* calls. The impact also of *HTTP pipelining* as well as new future Internet *HTTP*-modified networking protocols like *SPDY* [1] and *HTTP Speed+Mobility* [23] that offer reduced latency through compression, multiplexing and prioritisation need to be assessed. Additionally, other issues such as excess buffering of packets may cause high latency and jitter [5], and this may have significant impact on network performance, which might be a show-stopper for time-critical applications.

4.5 Conclusions

We have attempted to tackle four major critical questions that arise as key when technologies have to be selected and used to implement an SOA-based distributed large-scale process monitoring and control system. After compiling and assessing a set of technologies, a subset of them has been selected and used by the IMC-AESOP consortium. It is important to call attention to the fact that the selected technologies are either already available from open-source sites or are still under development by some of the IMC-AESOP technology-provider partners.

Following the assessment of the prototype implementations, which refine the technology evaluation and investigate other challenges in implementing SOA-based cross-domain infrastructures, e.g. cloud of services generated from the virtualisation of different systems like manufacturing, smart grid, transportation, etc., the experimentation results have shown that technological choices made are quite promising for next-generation SCADA/DCS systems:

- For real-time SOA, EXI, the binary XML format makes a lot of sense for wireless interconnections, in particular for CoAP-based data exchanges even though CoAP does not support true real-time capabilities in the current design. The benefit of EXI compression is also less obvious for wired SOAP-based Web services.
- Proposing a dual OPC-UA/DPWS stack can facilitate the management of large-scale distributed systems by building a bridge between industrial automation and IT worlds.
- An SOA middleware like the proposed distributed service bus can ease the integration and interoperability of heterogeneous technologies in the plant.

Acknowledgments The authors thank the European Commission for their support, and the partners of the EU FP7 project IMC-AESOP (http://www.imc-aesop.eu) for fruitful discussions.

References

1. Belshe M, Peon R (2012) SPDY protocol. IETF internet-draft. http://www.tools.ietf.org/html/draft-mbelshe-httpbis-spdy-00
2. Bormann C, Castellani AP, Shelby Z (2012) Coap: an application protocol for billions of tiny internet nodes. IEEE Internet Comput 16(2):62–67. http://doi.ieeecomputersociety.org/10.1109/MIC.2012.29
3. Castellani A, Gheda M, Bui N, Rossi M, Zorzi M (2011) Web Services for the internet of things through CoAP and EXI. In: IEEE international conference on communications workshops (ICC), 2011
4. Fojcik M, Folkert K (2012) Introduction to OPC-UA performance. In: Kwiecień A, Gaj P, Stera P (eds) Computer networks, communications in computer and information science. Springer, Berlin, vol 291, pp 261–270. doi:10.1007/978-3-642-31217-528. http://www.dx.doi.org/10.1007/978-3-642-31217-5_28
5. Gettys J, Nichols K (2012) Bufferbloat: dark buffers in the internet. Commun ACM 55(1):57–65. doi:10.1145/2063176.2063196. http://www.doi.acm.org/10.1145/2063176.2063196
6. Hilbrich R (2010) An evaluation of the performance of dpws on embedded devices in a body area network. In: IEEE 24th international conference on advanced information networking and applications workshops (WAINA), 2010, pp 520–525. doi:10.1109/WAINA.2010.93
7. Izaguirre M, Lobov A, Lastra J (2011) Opc-ua and dpws interoperability for factory floor monitoring using complex event processing. In: 9th IEEE international conference on industrial informatics (INDIN), 2011, pp 205–211. doi:10.1109/INDIN.2011.6034874
8. Jammes F, Bony B, Nappey P, Colombo AW, Delsing J, Eliasson J, Kyusakov R, Karnouskos S, Stluka P, Tilly M (2012) Technologies for SOA-based distributed large scale process monitoring and control systems. In: 38th annual conference of the IEEE industrial electronics society (IECON 2012), Montréal, Canada
9. Jennings C, Shelby Z, Arkko J (2013) Media types for sensor markup language (SENML). Technical report, IETF Secretariat. http://tools.ietf.org/html/draft-jennings-senml-10

10. Karnouskos S, Colombo AW (2011) Architecting the next generation of service-based SCADA/DCS system of systems. In: 37th annual conference of the IEEE industrial electronics society (IECON 2011), Melbourne, Australia
11. Karnouskos S, Somlev V (2013) Performance assessment of integration in the cloud of things via web services. In: IEEE international conference on industrial technology (ICIT 2013), Cape Town, South Africa
12. Karnouskos S, Colombo AW, Jammes F, Delsing J, Bangemann T (2010) Towards an architecture for service-oriented process monitoring and control. In: 36th annual conference of the IEEE industrial electronics society (IECON 2010), Phoenix, AZ
13. Karnouskos S, Colombo AW, Bangemann T, Manninen K, Camp R, Tilly M, Stluka P, Jammes F, Delsing J, Eliasson J (2012) A SOA-based architecture for empowering future collaborative cloud-based industrial automation. In: 38th annual conference of the IEEE industrial electronics society (IECON 2012), Montréal, Canada
14. Kyusakov R, Eliasson J, Delsing J (2011) Efficient structured data processing for web service enabled shop floor devices. In: IEEE international symposium on industrial electronics (ISIE), 2011, pp 1716–1721. doi:10.1109/ISIE.2011.5984320
15. Luckham DC (2001) The power of events: an introduction to complex event processing in distributed enterprise systems. Addison-Wesley Longman Publishing Co., Inc, Boston, MA, USA
16. Mahnke W, Leitner SH, Damm M (2009) OPC unified architecture. Springer, Heidelberg. ISBN 978-3-540-68899-0
17. Moritz G, Timmermann D, Stoll R, Golatowski F (2010a) Encoding and compression for the devices profile for web services. In: 24th IEEE international conference on advanced information networking and applications workshops, WAINA 2010, Perth, Australia
18. Moritz G, Zeeb E, Prüter S, Golatowski F, Timmermann D, Stoll R (2010b) Devices profile for web services and the REST. In: 8th international conference on industrial informatics (INDIN), Osaka, Japan
19. Moritz G, Golatowski F, Timmermann D (2011) A lightweight SOAP over CoAP transport binding for resource constraint networks. In: IEEE 8th international conference on mobile adhoc and sensor systems (MASS), 2011, pp 861–866. doi:10.1109/MASS.2011.101
20. OCG (2007) Sensor model language (SensorML) implementation specification. http://www.opengeospatial.org/standards/sensorml
21. Shelby Z (2010) Embedded web services. Wirel Commun 17(6):52–57. doi:10.1109/MWC.2010.5675778. http://dx.doi.org/10.1109/MWC.2010.5675778
22. Tilly M, Reiff-Marganiec S (2011) Matching customer requests to service offerings in real-time. In: Proceedings of the 2011 ACM symposium on applied computing, ACM, NY, USA, SAC '11, pp 456–461. doi:10.1145/1982185.1982285. http://www.doi.acm.org/10.1145/1982185.1982285
23. Trace R, Foresti A, Singhal S, Mazahir O, Nielsen HF, Raymor B, Rao R, Montenegro G (2012) HTTP speed+mobility. IETF internet-draft. http://www.tools.ietf.org/html/draft-montenegro-httpbis-speed-mobility-02
24. Trnka P, Kodet P, Havlena V (2012) OPC-UA information model for large-scale process control applications. In: IECON 2012—38th annual conference on IEEE industrial electronics society, pp 5793–5798. doi:10.1109/IECON.2012.6389038

Chapter 5
Migration of SCADA/DCS Systems to the SOA Cloud

Jerker Delsing, Oscar Carlsson, Fredrik Arrigucci, Thomas Bangemann, Christian Hübner, Armando W. Colombo, Philippe Nappey, Bernard Bony, Stamatis Karnouskos, Johan Nessaether and Rumen Kyusakov

Abstract As process control and monitoring systems based on a Service-Oriented Architecture (SOA) are maturing, the need increases for a systematic approach to migrate systems. The legacy systems are traditionally based on a strict hierarchy and in order to gradually allow additional cross-layer interaction, the migration procedure needs to consider both—functionality and architecture of the legacy system. The migration procedure proposed here aims to preserve the functional integration, organize the SOA cloud through grouping of devices, and maintain the performance aspects such as real-time control throughout the whole migration procedure.

J. Delsing (✉) · R. Kyusakov
Luleå University of Technology, Luleå, Sweden
e-mail: jerker.delsing@ltu.se

R. Kyusakov
e-mail: rumen.kyusakov@ltu.se

O. Carlsson · J. Nessaether
Midroc Electro AB, Stockholm, Sweden
e-mail: oscar.carlsson@midroc.com

J. Nessaether
e-mail: johan.nessaether@midroc.com

F. Arrigucci
Midroc Electro AB, Malmö, Sweden
e-mail: fredrik.arrigucci@midroc.com

T. Bangemann · C. Hübner
ifak, Magdeburg, Germany
e-mail: thomas.bangemann@ifak.eu

C. Hübner
e-mail: christian.huebner@ifak.eu

A. W. Colombo et al. (eds.), *Industrial Cloud-Based Cyber-Physical Systems*,
DOI: 10.1007/978-3-319-05624-1_5, © Springer International Publishing Switzerland 2014

5.1 Introduction

In order to include Service-Oriented Architecture (SOA) in the continuous evolution of control and monitoring systems there is a need for a strategy for successive migration of a legacy system into a complete SOA plant. This chapter provides a discussion on how to migrate a system to a Service-Oriented Architecture and how a migration can be expected to affect operations.

This process of migrating from a legacy process control and monitoring system to a Service-Oriented Architecture supports focus on the functionality and the loose coupling of heterogeneous systems to fit dynamic business needs. The legacy systems typically have proprietary protocols and interfaces resulting in vendor lock-ins and possibly site-specific solutions; however, with SOA these systems can be wrapped, extended or replaced and integrated in a modern infrastructure.

There is a considerable need to meet various migration requirements for small as well as large-scale investments, projects and upgrades of a process control and monitoring system. Here, the focus is on migration of large distributed automation systems. The migration towards new functionality, new technology as well as new systems, is risky and therefore the risks of downtime, poor performance and even failure to train personnel must be eliminated. Structured use of risk analysis facilitates the evaluation of different migrations paths.

The migration strategy has its starting point in the business needs, and ideally makes it possible to migrate from a legacy to new system seamlessly without noticeable interruptions at shop floor and business levels. A migration plan for the pertinent plant, should be compiled, and this has to be validated against global migration plans in order to assure that there are no direct interdependencies with other systems (local and enterprise wide).

It is important to evaluate the migration afterwards and question whether the requirements are fulfilled. Therefore, the requirements must be quantifiable and measurable. For example, in order to minimise negative impact of the migration enterprises need measurable requirements for effects like downtime, control

---

A. W. Colombo
Schneider Electric, Marktheidenfeld, Germany
e-mail: armando.colombo@schneider-electric.com

A. W. Colombo
University of Applied Sciences Emden/Leer, Emden, Germany
e-mail: awcolombo@technik-emden.de

P. Nappey · B. Bony
Schneider Electric, Grenoble, France
e-mail: philippe.nappey@schneider-electric.com

B. Bony
e-mail: bernard.bony@schneider-electric.com

S. Karnouskos
SAP, Karlsruhe, Germany
e-mail: stamatis.karnouskos@sap.com

problems, costs, interoperability, performance and possibly personnel training. Generic migration strategies, where the different paths and steps are discussed in some more detail, are described hereunder. For this generic migration, the proposed methodology will be developed providing general directions to implement an efficient and low-risk transition from an old system to a SOA-based monitoring and control system in process industry environment.

The legacy systems are typically implemented following the hierarchically organised 5-level model as defined within the ISA-95 / IEC 62264 standard (http://www.isa-95.com). Operations, defined by that standard, are inherent to established production management systems [7]. In this context, concepts for integrating legacy systems, specifically on lower levels, into Service-Oriented Architecture-based systems can be seen as business enablers to take the customer from where she/he is today [9] into the future.

Several provider of today's enterprise systems, Level 4 in the ISA-95 architecture (please refer to Chap. 2) already support service-driven interaction, e.g. via Web services. Service-Oriented Architecture is an approach used at this level. Services are also used for integration between Level 3 and Level 4 systems, available on the market. OPC-UA [13] is a technology spreading-up to be used. PLCopen in close cooperation with OPC Foundation, defined a OPC-UA Information Model for IEC 61131-3. A mapping of the IEC 61131-3 software model to the OPC-UA information model, leading to a standard way how OPC-UA server-based controllers expose data structures and function blocks to OPC-UA clients like HMIs was defined [1]. OPC-UA relies on Web service-based communication. Such activities can be seen as attempts to move towards the use of common technologies across different levels of production systems.

By abstracting from the actual underlying hardware and communication-driven interaction and focusing on the information available via services, the complete system is managed and controlled by service-driven interactions. Services can be dynamically discovered, combined and integrated in mash-up applications. By accessing the isolated information and making the relevant correlations, business services could evolve; acquire not only a detailed view of the interworking of their processes but also take real-time feedback from the real physical-domain services and flexibly interact with them.

The novelty of migrating from a legacy process control system into a SOA, is to do that in a structured way, gradually upgrade highly integrated and vendor-locked standards into a more open structure while maintaining the functionality. The challenges of stepwise migration of a highly integrated vendor-locked DCS and/or SCADA are discussed. From here the necessary migration technology and procedures are proposed. The critical migration technology proposed is based on the mediator concept (as described in Chap. 2). The migration procedure proposed is based on a functionality perspective and comprises four steps: initiation, configuration, data processing and control execution. It is argued that these steps are necessary for the successful migration of DCS and SCADA functionality into a service-based automation cloud.

5.2 Challenges in Migrating Industrial Process Control Systems

Today's control systems, as used in process or manufacturing automation, are typically structured in an hierarchical manner as illustrated in Chap. 2 in Fig. 2.1.

IEC 62264 (or originally ISA-95) [7] is the international standard for the integration of enterprise and control systems, developed to provide a model that end users, integrators and vendors can use when integrating new applications in the enterprise. The model helps to define boundaries between the different levels of a typical industrial enterprise. ISA-95/IEC 62264 define five levels. For each of these five levels certain problems and challenges become eminent when considering their implementation using a SOA-based approach.

Whereas Level 0 is dedicated to the process to be controlled itself, Level 1 connects the control systems to the process by sensors and actuators. Through the sensors the control system can receive information about the process and then regulating the process through the actuators. Sensors convert temperature, pressure, speed, position etc. into either digital or analogue signals. The opposite is done by actuators. Including not only valves but also motors and motor equipment such as frequency converters in actuators, it can be said that the level of installed intelligence varies very much. Legacy implementations use a scan-based approach reading and writing data from/to sensors/actuators. Which differs fundamentally to the event- based nature of a SOA approach [8, 11]. Migration on Level 1 has to some extent been described [3] with focus on transition from scan-based to SOA event-based communication when it comes to analogue signals.

At Level 3, operational management of the production is done, where Manufacturing Execution Systems (MES) provide multiple information and production management capabilities. In the context of control hierarchy, however, its main function is the plant-wide production planning and scheduling. In a continuous process plant, the results of scheduling are used as production targets for individual shifts, and consequently, translated by engineers and operators into individual set points and limits. Level 3 integrates information about production and plant economics and provides detailed overview about the plant performance. If the production is straight forward with few articles and small production site, a dedicated Level 3 system might not bring added value. Some typical MES/MIS functionality is instead put in Level 2 and/or in the ERP-system (Level 4). At Level 4, typically Enterprise Resource Planning Systems (ERP) are installed for strategic planning of the overall plant operation according to business targets. Migration into SOA at Level 3 and 4 does not differ significantly for factory automation and process control systems [2].

At Level 2 there are some non-resolved challenges of migration when it comes to the process industry. Distributed monitoring and control enables plant supervisory control. The Distributed Control System (DCS) of a large process plant is usually highly integrated compared with a SCADA solution which is standard in factory automation. The SCADA is a supervisory system for HMI and data acquisition and the system communicates through open standard protocols with subordinated

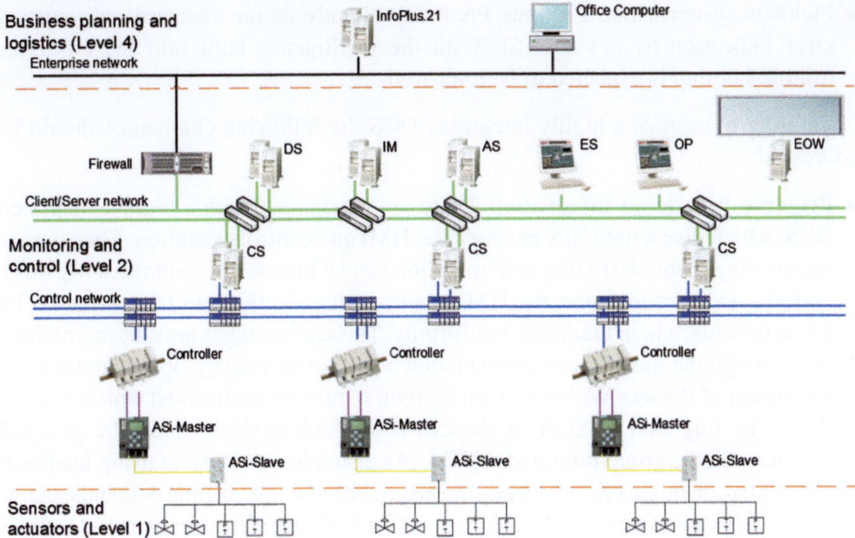

Fig. 5.1 Legacy system architecture

PLCs. The PLCs in the SCADA solution are autonomous compared to their counterpart, which sometimes are referred to as controllers, in the DCS. In this paper the process control system is defined as a DCS including HMI workstations, controllers, engineering station and servers all linked by a network infrastructure. A DCS is truly 'distributed' with various tasks being carried out in widely dispersed devices. Migration of Level 2 functionality in the form of a DCS exhibits challenges when it comes to co-habitation between legacy and SOA as well as the migration of the control execution [8, 11]. Here, the DCS is exemplified by a server/client- based system as depicted in Fig. 5.1, which is a common topology.

When migrating the DCS into SOA there are certain requirements based on expectations from business, technical and personnel perspectives:

- The new architecture and the migration strategy must assure the same level of reliability and availability as the legacy system.
- The migration procedure must not induce any increased risk for staff, equipment or process reliability and availability.
- After the migration the plant must still provide the same or a better process, extended service life of plant (process equipment, e.g. pumps, vessels, valves), adequate information and alarms depending on department and personnel skill and improved vertical (cross-layer) communication with more information available at plant- wide level.
- Dynamic changes and reorganisation is expected to be supported, on a continuously running system.
- To handle the co-habitation between the legacy system and the SOA during the migration phase, the SOA solution must support wrapping of legacy subsystems.

- Fieldbus systems, like Profibus PA today already define standardised ways of error indication by devices [5]. With the intelligence built into SOA devices troubleshooting is expected to be improved.

In order to migrate a highly integrated DCS the following challenges should be addressed:

- **Preserve functional integration** There are advantages with a highly integrated DCS, which give a tight link between the HMI and control execution. Thus design engineering, commissioning and operation can be pursued in a significantly more uniform way. For instance, the HMI and control execution can be configured by the same tools, which facilitates conformity. These advantages must be maintained even though the integration is broken down and substituted by open standards.
- **Grouping of devices** Within a given system, it must be determined which devices should be migrated to SOA as devices and which devices should be grouped together and the group migrated to SOA. As example a subsystem using feedback and regulation might require legacy interfaces because of real-time demands, therefore such group of devices should be given an SOA interface for the group using a Mediator and not at device level. This part of the system may be handled as 'black box'
- **Preserve real-time control** The real-time control execution, which in the legacy system is secured in the controllers, must be preserved.

5.3 Functional Aspects Identified in a DCS

To support the preservation of key functionality during and after the migration certain functional aspects of a generalised Distributed Control System have been identified. In this section a short description is presented for each aspect in order to provide a frame of reference for the migration approaches presented in the following section.

- *Local control loop* The function of a Local control loop refers to the low-level automated control that regulates a certain part of the plant process, with a relatively low number of actuators and sensors. The control may be continuous or discreet and may use analogue as well as digital actuators and sensors. In many cases the control will require low latency and short sample times, resulting in high bandwidth.
- *Distributed control* This refers to all forms of control where parts of the control loop are located far away from each other, geographically or architecturally, meaning that the control cannot be executed by a single device (controller) with direct access to both sensors and actuators.
- *Supervisory control* This form of control is often executed at a higher level based on information from more than one subsystem and is usually much slower than the Local control loop. Often the Supervisory control has no direct access to sensors

or actuators but uses aggregated process values as input and actuates through changing the set point of a Local control loop.

- *System aggregation* Low-level devices and subsystems are often presented in an aggregated form to higher level systems in order to show an understandable overview of the system to operators, engineers and others working with the system.
- *Inter-protocol communication* As different levels of the DCS use different communication standards and protocols all communication between components that are not connected to the same network type and in the same or neighbouring segment need to pass the information through one or more other components. These other components must therefore be able to interpret or translate the information between the different standards and protocols. The effort needed for this kind of communication varies greatly depending on the standards and protocols involved.
- *Data acquisition, display and storage* Process and system data gathered at all levels of the DCS must ultimately be made available to operators and other connected systems. The availability of correct data is vital to both—operators and management in order to optimise performance and analyse anomalies. In some cases, historical data storage is integrated in the DCS but even in these cases the functionality is not an integral part of the DCS functionality and can be treated as a peripheral system.
- *Alarms and warnings* All systems have some way of indicating process anomalies to the personnel working with the process. In a well-developed DCS, there are many functions related to alarms and warnings that allow distribution of information to the appropriate staff and several modes of suppression and acknowledgement of alarms and warnings.
- *Emergency stop* The Emergency stop is a vital part of most process control systems, often regulated by national laws and regulations. In a large process plant, the emergency stop may be much more complex than simply shutting off the power to all components as this may cause situations where a build-up of heat or pressure, or a chemical reaction would cause a greater disaster than to keep the plant running. It is important that a process control system is able to execute a reliable shut-down procedure even in unexpected situations.
- *Operator manual override* At most plants it is required that the operator can control parts of the system manually, via an HMI, to handle irregular or unexpected situations. This may be to support maintenance operations where systems are disconnected in a controlled manner or when the operator has to handle unexpected faults in the process or in the automation system.
- *Operator configuration* Most operator stations allow changing of some parameters in the system such as plant or system operation mode, or control set points for subsystems based on information not available in the automation system.
- *User management and Security* As many parts of a DCS are interconnected and there are many people with different roles that work with a DCS, it is important that each person is presented with a level of information that is sufficient and relevant for their role. In order to limit human errors as well as malicious actions it is important that all personnel are authenticated for the role in which they are

allowed to access the system. The authentication may not always be limited to the software but may instead consist of limiting physical access to certain areas or stations.

5.4 Migration of Functionality

In order to ensure and support the preservation of functionality throughout the migration process each functional aspect identified in a DCS have been analysed and for each aspect an example is presented on how the migrated system could provide the functionality in question. These examples are not necessarily the only or the optimal implementations of the functionality but they should provide sufficient example covering the complete DCS.

5.4.1 Local Control Loop

At the level of local control loops, the main benefit of applying the SOA communication infrastructure is the richer set of diagnostic and monitoring information that can be delivered and easily integrated into the SCADA systems. By using standard service protocols for the sensor and actuator data delivery, the provisioning stage can be automated to a higher degree than what is possible with the current approaches. Also modifications and upgrades to the system are better supported by using modular, loosely coupled services with support for event-based interactions and resource discovery. As part of the IMC-AESOP project two main approaches are available to migrate the existing control loops to SOA-based solutions proposed by the project:

- For control loops with low real-time requirements (loop times around 100 ms or higher), the IMC-AESOP services 'Sensory data acquisition' and 'Actuator output' can be deployed directly to the embedded sensor/actuator devices. By the use of EXI and CoAP technologies, it is possible to provide extensive and non-intrusive diagnostics and monitoring information through wireless links. In many scenarios, the achieved efficiency is envisioned to support even the communication of process values via low-bandwidth wireless solutions. Legacy devices supporting firmware updates can be migrated directly to this architecture. For closed black box devices the IMC-AESOP services 'Gateway' and 'Service Mediator' are required to provide SOA interface and protocol mapping.
- For control loops with strict timing requirements and short loop times (below 100 ms) the direct deployment of 'Sensory data acquisition' and 'Actuator output' requires deterministic and high-bandwidth PHY/MAC layers such as Industrial Ethernet solutions. For low-bandwidth links, e.g. (Wireless) HART, would likely require gateway/mediator wrapping to migrate the low-level real-time protocols

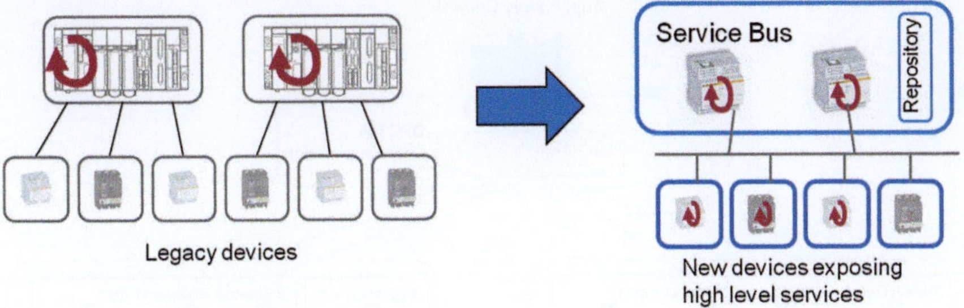

Fig. 5.2 Distributed control with a Service Bus middleware

used for the loops with a SOA-ready interface. Thus, simple and time-critical sensors/actuators part of real-time control loops are not migrated to SOA but rather wrapped on a higher level.

5.4.2 Distributed Control

A service architecture supports the distribution of the treatments on several systems or devices. As far as it is possible the control is located at the lowest level so that the treatments can be more appropriate due to the knowledge of the local context. Moreover, the amount of data needed to communicate to the upper levels can be reduced.

Intelligence of the control is pushed down in the devices so that treatments remaining at controller level may be performed in other kinds of devices than controllers, like for example, network infrastructure devices. Part of the control can be temporarily disconnected without affecting the complete equipment, either for normal replacement or even for upgrading the functionalities. Configurations containing the control logic are stored in a central repository so that exchange of devices is possible without manual reprogramming. A Service Bus middleware can typically support such a decentralised control:

- The components of the Service Bus may be physically distributed on:
  - Existing devices of the system (case of the Fig. 5.2)
  - Existing infrastructure devices like gateways
  - Dedicated devices

- Some of the devices have their own local logic so that they can expose high level services.
- Other devices cannot expose such SOA services. They are either legacy devices or small devices that do not support local logic. Thanks to the Service Bus these devices can nevertheless interoperate in the system.

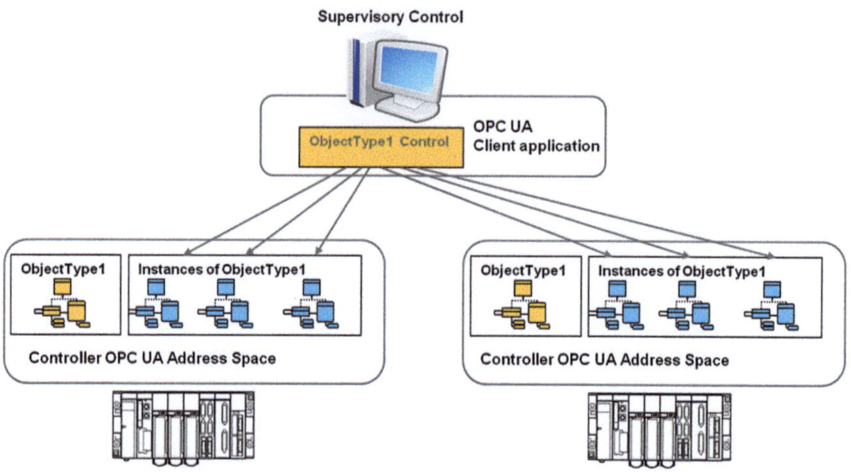

Fig. 5.3 Programming against object types with OPC-UA

- The remaining logic required to make the global control of the system is distributed within the Service Bus, i.e. in this example on the two devices supporting a Service Bus component.

5.4.3 Supervisory Control

In a SOA approach, devices can expose directly their data to the other systems at different levels; there is no more a hierarchical structure where device data are first collected by controllers which then feed the supervisory control system. The visibility of the devices is then improved without additional workload. Maintenance and evolutions of the supervisory application are also decoupled from other underlying systems like controllers or OPC servers.

Supervisory control systems can also propose a richer interface while their development is easier thanks to the usage of tools understanding the standard interfaces exposed by the controllers and the devices. These interfaces are typically described through WSDL files.

OPC-UA provides additionally a feature known as programming against type definitions (see Fig. 5.3 below). The principle is that an OPC-UA server supports the definition of complex object types which can be recognised by a client application like a supervisory control. In the server address space both the object type and the object instances are exposed. The supervisory control either already knows the object types exposed by the server or discovers them during the engineering phase. In both cases, the treatments concerning each object instance is programmed only once due to the knowledge of the object type. In this way, supervisory control applications can be quickly developed with libraries of components corresponding to standard object types.

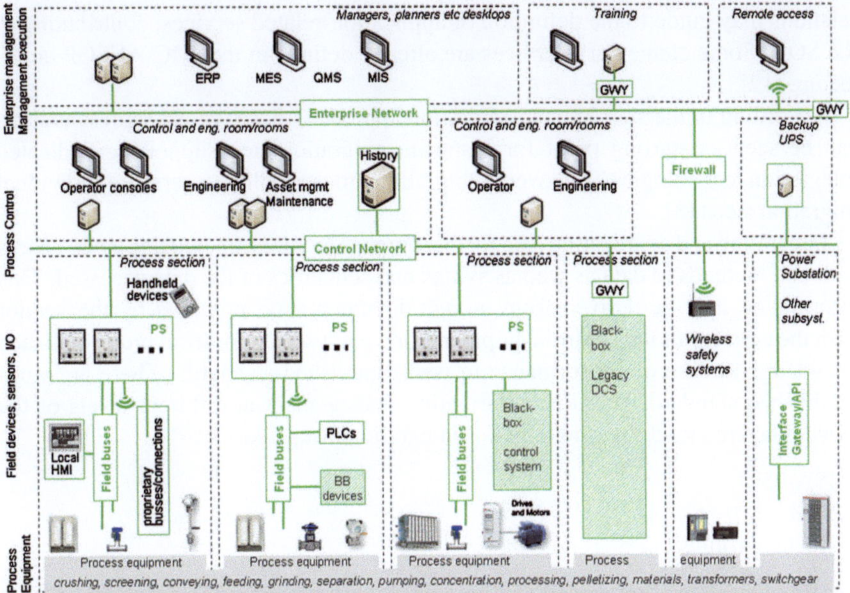

Fig. 5.4 General architecture of a process control system

5.4.4 System Aggregation

As indicated in Fig. 5.4, process plants are separated into several sections. Depending on the nature of the process represented by a section, control can be realized in an encapsulated, but coordinated by master control, manner. This is even more the case in batch applications than in continuous processes. Batch control is a more flexible way for mastering market demands of producing small quantities of changing products (chemical, petrochemical, medical, etc.) at the same production site. Here production equipment like boiler, heat exchanger, distillation colon, or alike, are dynamically combined and controlled according to recipe needs. Support functions like air compression for auxiliary energy provision or cooling aggregates are normally built as package unit also having own controls.

As it can be seen, today's classical process plants, and associated automation systems are already, even if partially, characterised by

- Aggregation of information dedicated to specific plant sections
- Individual engineering and control of those sections (black boxes)
- Hierarchical engineering concepts for overall/master control
- Supervision down to black box level.

Additionally, one can start from the process level to identify plant sections, e.g. performing individual control loops or contributing information to dedicated aspects (like Maintenance) of a plant view, to define data related to each other. Those relations may guide to the definition of application-related services, contributing to the SOA. Some elementary services are already defined in the IMC-AESOP architecture [12].

According to the step-by-step Migration approach, those typical representations can be seen as starting point for a specific migration step supporting dedicated integration technologies. The overall migration process will be a series of individual migration steps [3].

Integrating and aggregating data for that purpose requires, knowledge about access path and methods to data as well as syntax and semantics of the data accessed. This information may be derived from project documentation provided by the vendor with the delivery of the control equipment for a plant section or from pre-established knowledge in case of conformance to well-established standards. There are well-established standards targeting information management at different levels of the ISA-95 layered model, exploitable for integration tasks, like:

- ISA-95
- S 88
- Device Profiles.

Use of standard conformant equipment is highly recommendable, as they ensure reimbursement of investments. Afterwards, a summary of a concept for aggregation of data and definition of services at Level 1 and 2 is given continuing work started in SOCRADES [2, 14].

A Gateway and Mediator concept [10] has been introduced as being suitable for realising integration tasks within the IMC-AESOP framework. This concept supports representing single resources (like a legacy device) to a SOA-based environment as well as aggregating and mediating data from a single or multiple resources.

5.4.5 Inter-protocol Communication

Interoperability of applications requires fundamental communication capabilities, even if applications are running on inhomogeneous communication platforms. That, in fact, is the usual case for integration tasks. Two, or more, communication channels have to be mapped to each other considering the different characteristics at all protocol layers. There are different approaches known from literature:

- Bridge
- Gateway
- Router.

Considering introduction of SOA into process control environment, one will be faced with integration tasks of different type of communication (4 . . . 20 mA standard-wired signals, HART protocol, fieldbus protocols like Profibus PA and others). All to be mapped to a single protocol, as agreed to be used for communication within the SOA.

Within the IMC-AESOP approach Gateway or Mediator concepts are used for protocol mapping, covering interfacing of different protocols, interpreting syntax and semantics of data operated at each communication channel (possibly in a different way) and mapping data to an internal data model of the integration components. The Web Server (interface to SOA) accesses the internal data model and maps the data to an appropriate Web service, conformant to the IMC-AESOP architecture definition [12].

Configuring this mapping is a multistep approach, while doing configuration for each of the individual communication channels, instantiating an internal object model, representing the targeted view of the underlying system, and defining the mapping roles to the Web services. Knowledge is needed for all the protocols and applications targeted.

5.4.6 Data Acquisition, Display and Storage

Data acquisition, in terms of the current state of the art has many possible solutions and implementation, most commonly using a PLC or some sort of RTU connected to a fieldbus to transfer data as required. In terms of the IMC-AESOP architecture the main objective is to change, or migrate from this kind of traditional systems to smart embedded devices capable of both acquiring the necessary data and encapsulating it in Web services that can later be consumed by any interested party. An example of this migration could be taken from Use Case 2 Oil Lubrication of the IMC-AESOP project.

At the lowest level this use case requires computers capable of calculating flow rates from positive displacement flow metres. These volumetric flow metres generate pulses at frequencies ranging from 1–500 Hz depending on the model being used. Any conventional PLC or RTU unit has inputs that can detect a frequency of roughly 50 Hz. While this is good enough for certain flow metres it is not nearly enough to cover the whole range of possibilities. There are two possible solutions to this migration problem:

1. One possible solution is to use a legacy flow computer with legacy communication capabilities, i.e. modbus. This would enable the flow computer to do the high frequency calculations necessary and transfer the data to a modbus register that could be read by any WS-capable device. The data would then be processed from pulses into flow rate and encapsulated into a WS-Event, or message depending on whatever the requirements are.

2. Another possible solution is to have a fast counter card or specialised inputs integrated to a WS-Capable device. This would imply that the device would both have to be capable of counting, pre-processing and calculating flow rates without any external help. Then it would only be a matter of encapsulating the data in WS form in order to make it available to any interested parties.

Whichever solution is chosen however, it is important to keep in mind that legacy flow metering computers, while limited, have well-defined, well-tested algorithms that calculate flow rate. In the case of migration it is necessary to evaluate whether accuracy or scalability and easy access are more important. The common in both solutions however, is the need for WS encapsulation, which would imply exactly the same work in both cases. It would be necessary to design the corresponding WSDL file so that the device capturing the information could be discovered and subscribed to, Although this might depend on how the WS-Capable device is designed to work.

5.4.7 Alarms and Warnings

Alarms can be raised at different levels, either directly by the devices or by upper level systems, processing various information coming from one or several sources.

Additionally to the definition of standardised interfaces defining the content of the alarms, an SOA approach proposes communication mechanisms insuring that:

- The right information will reach the right person in the plant and with an appropriate level of details
- The communication network of the plant will not be overcrowded by useless data.

These two goals are achieved by filtering and routing mechanisms, implemented typically by a Complex Event Processing (CEP) technology as investigated in the IMC-AESOP Project.

For the end user, the benefit of a SOA approach is that he will receive only the needed alarms and warnings. The content of the alarms will be filtered depending on the user who is logged into the system, giving the information just required for the actions of the user. For example, an operator will be informed that the process is stopped without any further detail while a maintenance team will receive details about the machine breakdown.

5.4.8 Emergency Stop

Detection of abnormal conditions requiring an emergency stop can be performed at various levels. Additionally to emergency stop buttons at shop floor level, the events raised in the different layers of the system can indirectly inform the operators of critical alarms, typically within the supervisory control system. Moreover, complex event processing systems can correlate the information coming from different sources located in any location of the system in order to raise such emergency alarms.

Once the operator has pressed physically a shop floor button or has selected the emergency stop in an HMI, the equipment must shutdown in a controlled manner which depends on the exact state and topology of the system. The agility of a SOA infrastructure allows managing several shutdown strategies depending on the various emergency conditions as well as adapting these strategies all along the life cycle of the equipment.

In some context, typically for regulation purpose, the shutdown of the equipment must be done in a given time frame and with a precise sequence of operations. In those cases, safety protocols solutions must be used to manage these particular constraints. There are currently different add-ons existing for classical fieldbuses but for the envisioned systems where IP protocol over Ethernet is largely used, safety solutions based on Ethernet must be carefully considered.

5.4.9 Operator Manual Override

The devices expose standardised interfaces so that a unique or at least a limited set of tools can be used by the operators for taking the control locally. Then the operators can be well trained and efficient, what is particularly important when an unexpected situation happens, which is a typical case where manual override is required.

The parts of the system where operators have overridden the automatic control must be easily pointed out in the upper levels applications, even if this is a scheduled maintenance where a part of the system is disconnected intentionally. SOA makes possible a direct connection between the upper level and the devices so that such critical information is easily available. Such information is used not only by the operator but also by the upper level applications to reconfigure themselves.

Thanks to the loose coupling of the SOA approach, most applications at level 2 or level 3 will continue interacting with the manually controlled part of the system without considering its operating mode. Only applications interested by the operating mode will be informed, typically by alarms and events mechanisms.

5.4.10 Operator Configuration

The devices expose standardised configuration services so that here also a limited set of tools can be used for local configuration. Then operators have to get a lot of different tools and to be trained for them. The changes made in the device configuration must be then pushed to the configuration repository in order that after replacement of a device, the same configuration can be downloaded to the new one. Different strategies can be used here, either the operator decides explicitly that the new settings are valid and initiates the backup manually, or the device configuration may be compared periodically to the reference, which is updated if the actual device configuration is different but valid.

The Fig. 5.5 below describes a system where a standard service DeviceManagement is supported by an IMC-AESOP device. A local configuration tool can be used to perform the following actions:

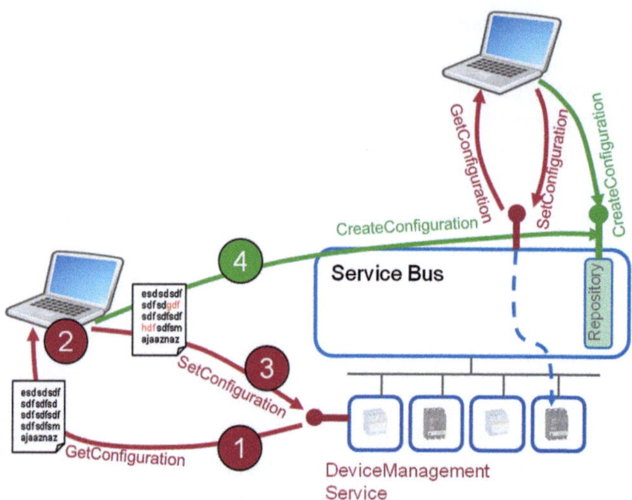

Fig. 5.5 Operator local configuration with a Service Bus

1. Get the current configuration of the device. The response of the GetConfiguration operation is defined with a very generic format. Virtually any kind of device configuration can be retrieved.
2. The operator edits the device configuration with the configuration tool HMI.
3. The tool uploads the new configuration in the device (SetConfiguration operation).
4. Optionally, the new configuration is pushed in the configuration repository. This repository will be used in particular in case of device replacement.

   Notes:

- In this example, the configuration repository is managed within the Service Bus introduced in Chap. 4
- The right side of the Fig. 5.4 demonstrates that the Service Bus can provide a service view also for legacy devices. It translates legacy protocols and legacy data formats so that it can expose the DeviceManagement service on behalf of the legacy devices.

5.4.11 User Management and Security

Up to now, a predominant behaviour was to have locally authenticated users (e.g. on-device or department) and devices (if at all). However, this practice created 'islands' within the infrastructure that were difficult to be controlled, e.g. if they adhere to the corporate policies, and are costly to maintain. However in the IMC-AESOP vision, the security framework should be company-wide and the 'visibility' of devices in the cloud makes it easier to have a system-wide view. The migration, however, towards this infrastructure, will require a lengthy transition process and potentially significant effort to reassess security and risk relevant aspects, test configuration and impact, and move towards integrated management of both users and their rules.

5.5 Migration Procedure

Interfacing and integrating legacy and SOA components of a DCS/SCADA system will require some, for the purpose developed and/or adapted, technology. Such integration may be based on some kind of integration component like Gateway or Mediator. Such Gateway or Mediator has the task to bridge the communication from major standardised protocols used close to field applications today: HART communication, Profibus PA in combination with Profibus DP, Foundation Fieldbus, etc. These protocols follow specific characteristics. Some commonalities can be monitored like concepts for device descriptions or integration mechanisms into DCS (e.g. EDD, FDT, FDI). The same bridging task exists regarding communication to higher level, technologies related to Enterprise Application Integration (EAI) or Enterprise Service Bus (ESB) or OPC (OPC DA, OPC-UA) are used having their own characteristics and configuration rules.

The use of Gateways or Mediators is a well-proven concept for integrating/connecting and migrating devices, attached to different networks. It is used to transform protocols as well as syntax of data. Semantic integration is hard to achieve. Nevertheless it is possible to do transformation between data centric approaches, as typically followed by fieldbus concepts, and service-oriented, event centric, approaches.

The Mediator [10] concept used here is built on the basis of the Gateway concept by adding additional functionality. Originally meant to aggregate various data sources (e.g. databases, log files, etc.), the Mediators components evolved with the advent of Enterprise Service Buses (ESBs) [6]. Now a Mediator is used to aggregate various non WS-enabled devices or even services in SOA [10].

Using Mediators instead of a Gateways, provides the advantage of introducing some semantics or to do pre-processing of data coming from legacy networks, e.g. representing a package unit. Due to the diversity of data, or different aspects of interest, that different applications request different types (e.g. quality, quantity and granularity) of data, interface devices will normally be built as a combination of Gateway and Mediator. As it may also be applicable to integrate service-oriented sections (e.g. retro-fit of a plant section or replacement of a package unit) into existing systems, this Gateway and Mediator concept can be extended to represent services into data centric systems (today's legacy systems). Mediator as well as Gateway concepts, both are powerful means for integrating single legacy devices or legacy systems encapsulating 'isolated' functionalities.

Whereas the operational phase of a system will benefit from the functionalities described above from the beginning of the migration process, engineering will be

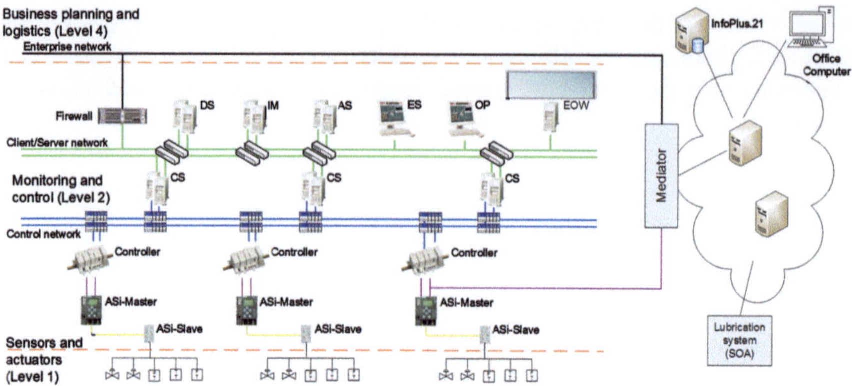

Fig. 5.6 DCS after the first step of migration

characterised by a stepwise approach, starting with defining services representing the legacy device or system, followed by separate engineering steps for the legacy part and the SOA-based part using those services defined. Specific configuration effort for the Mediator or Gateway itself is needed. It is advisable, that commissioning will also be done in a multiple step approach, starting at the isolated components followed by their integration into the overall system.

Considering the layout of a server/client-based SCADA/DCS a stepwise migration through four major steps is proposed. The four major steps may contain substeps and may be spread out over a long period of time but each major step should be completed before the following step is initiated. The four major steps suggested are:

- Initiation
- Configuration
- Data processing
- Control execution.

During the whole migration, the system will require one or more mediators to allow communication between the SOA components and the parts of the legacy system that not yet has been migrated. The propagation of the mediator and the growth of the SOA cloud are exemplary applied to the migration of the legacy SCADA/DCS presented in Fig. 5.1. Making emphasis in the DCS-part, the set of Figs. 5.6, 5.7, 5.8 and 5.9 shows the different results reached throughout the whole migration process.

5.5.1 Step 1: Initiation

The initial SOA 'cloud' needs some of the basic services presented in [12] in order to support basic communication and management of the cloud. Once the basic architecture is, the constructed first peripheral subsystems can be migrated and new compo-

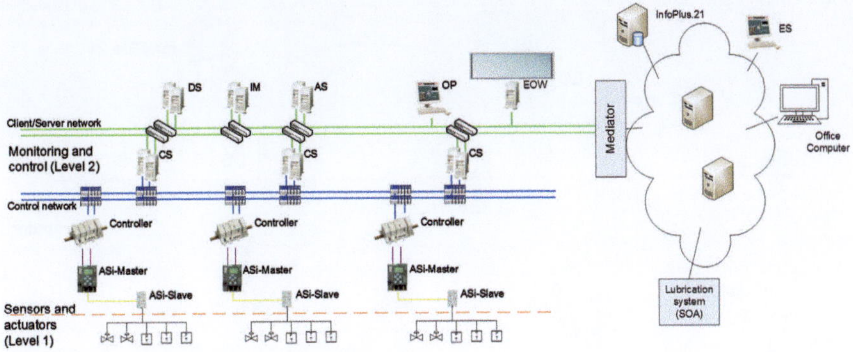

Fig. 5.7 DCS after the second step of migration

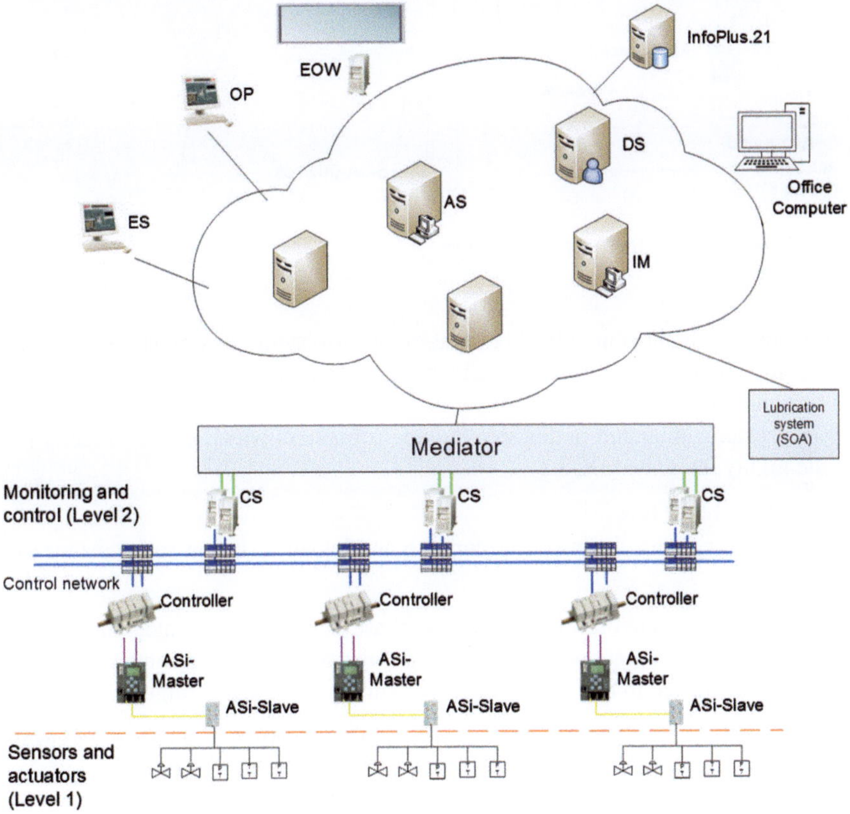

Fig. 5.8 DCS after the third step of migration

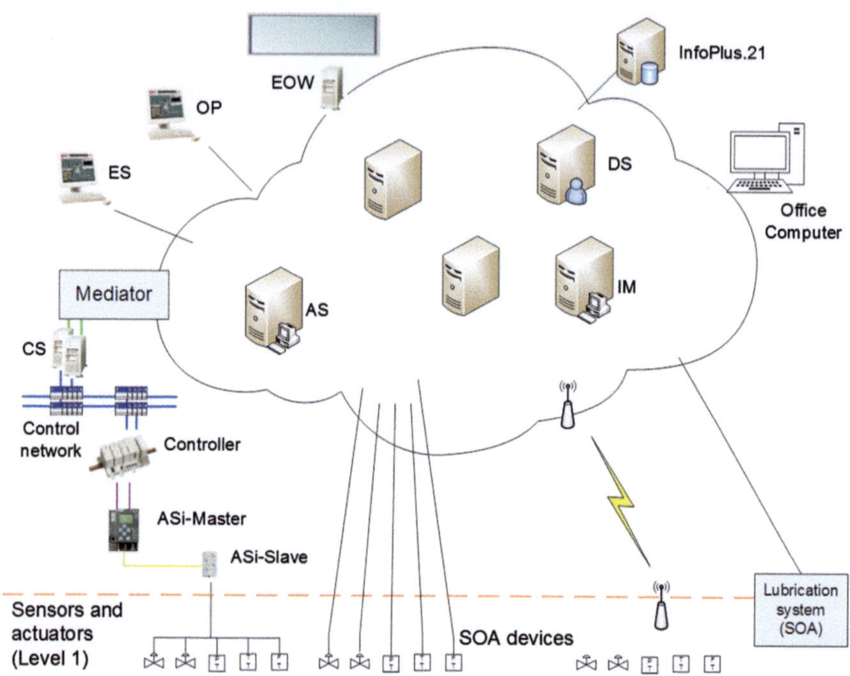

Fig. 5.9 DCS after the forth step of migration

nents can be integrated in SOA. In migration of subsystems, as well as integration of new components, some consideration must be made of the limitations of the mediator and its communication paths.

The systems migrated in this step include subsystem which are not directly part of the highly integrated DCS:

- Low-level black box
- High level systems for business planning and logistics such as maintenance systems.

Migration is limited to the operational phase of the systems integrated. Within that step, engineering is out of the scope of migration. An appropriate engineering approach, dedicated to this migration step, is doing multistep configuration:

- Configuration of every legacy system including the legacy interface within the mediator
- Configuring the SOA system
- Configuring the model mapping within the mediator.

Exploiting machine readable legacy configuration information would be helpful for every step. Today, configuration information is available through different technologies, e.g. GSD, EDD, paper documentation. This type of information is mostly available for single devices. Engineering stations take these information as input and generate system configuration information in proprietary formats.

At this point several parts of the functional aspects can be considered to be at least partially migrated. Most likely some of the Local control loop functionality is migrated. Inter-protocol communication is required both in the migrated and the traditional parts of the system and user management and security must be at least partially implemented in the SOA-system without compromising existing security or creating unnecessary obstacles for users or user administrators. System aggregation, emergency stop, alarms and warnings, operator manual override and operator configuration have all been implemented in the SOA-system to the extent required by the migrated subsystems, while the respective functionality in the traditional system is virtually untouched.

5.5.2 Step 2: Configuration

This is the first step where components that are heavily integrated in the DCS are migrated. The purpose of this step is to migrate parts of the DCS that do not require very short response times or the regular transport of large amounts of data. Please refer to Fig. 5.7. The majority of functions that qualify for this migration step are in some way concerned with configuration of different parts of the DCS. The point of origin for most, if not all, configuration is the Engineering Stations (ES) which is used for engineering and configuration of most parts of the DCS.

As the ES is migrated to SOA, this constitutes a major increase in the number of services the Mediator needs to supply to the SOA cloud as it must in addition to the operational data migrated in the first step represent configuration aspects of all legacy systems and devices not yet migrated, and allow configuration of all systems and devices. This means that configuration of low-level devices and control is done on the ES in a SOA environment using configuration services provided by the mediator, the configuration is then compiled by the mediator into their respective legacy formats and downloaded into the legacy controllers.

Configuration of HMI, Faceplates and associated systems is similarly done in SOA and converted by the mediator to a format that can be downloaded into the legacy Aspect servers and other legacy systems. The configuration of legacy devices from SOA might also require that the mediator is able to extract legacy designs and configurations that may be stored in aspect servers or controllers can be reused and modified by the SOA Engineering stations.

This approach may be combined with doing multistep configuration described in the former step.

As legacy systems usually do not provide sufficient meta-data, sufficient configuration information cannot necessarily be extracted by a Mediator from the installation (legacy systems). Consequently, for overall engineering a SOA engineering station should be able to import relevant configuration information of different legacy systems in addition to the limited capabilities provided by the Mediator itself. If such a tool would be available, one could design a mediator acting as configuration station for different legacy systems (compile configuration information into legacy formats) while receiving basic configuration information from the SOA engineering station.

As most of the functionality of everyday operation should be unaffected by the migration of engineering and configuration tools, only a few of the functional aspects are affected. Most notably there will be an increased need for inter-protocol communication and there may be a possibility to utilise more of the functionality described in Supervisory control. In addition, the migration of the Engineering station means that some additional parts of user management and security is migrated, but apart from those, most functional aspects should be similar to that those of the first step in the migration procedure.

5.5.3 Step 3: Data Processing

In this third step, the migration includes all components and/or subsystems that do not require short response time (millisecond range) not currently achievable by the SOA technology (refer to Fig. 5.8). This includes Operator Clients (OP) and Operator Overview Clients (EOW) as well as Aspect Servers (AS) and Information Management Servers (IM). As all points of user interaction with the system is now moved to SOA this means that the legacy Domain Servers (DS) are redundant. However, as user management and security needs to be available in SOA from the first step of the migration, there is probably no need for the Domain Servers in the SOA cloud, although the functionality can be considered to be migrated.

The migration the Operator Clients and the Aspect and Information Management Servers mean that the role of the mediator is once again fundamentally changed. In Step 3 of the migration, there is less of a need for a flexible mediator that can communicate with a lot of different legacy components, the new requirements are more concerned with a need to present large amounts of data available from legacy controllers to the migrated Operator Clients and other data processors and consumers. This activity is closely related to the purpose of the Connectivity Servers (CS) and it is suggested that the mediator in Step 3 is implemented as a new interface in the Connectivity Servers.

At this stage several operator-centric parts of the functionality are completely migrated. Most significantly Operator manual override and Operator configuration are fully migrated. All of Data acquisition, display and storage, except the first level of acquisition of data from the devices up to the controllers, are also migrated at this step. As the functionality for data acquisition is migrated some additional functionality for System aggregation might be required to present the data from underlying systems in the case where this is not sufficiently covered by the traditional systems. In addition all of the Alarms and warnings functionality, apart from some generation of alarms at the controller level, is migrated and so is most of User management and security.

5.5.4 Step 4: Control Execution

In the fourth and final step of migration the time has come to migrate the functionality traditionally provided by controllers (shown in Fig. 5.9). As control execution in the legacy system can be grouped together with several control functions in one controller, or in some cases spread out with different parts of a control function executed by more than one controller, it is of utmost importance that control execution is migrated function by function rather than controller by controller.

Depending on the performance requirements of each control function there may be a need for different strategies for different functions. In the cases where SOA compliant hardware is available for all functions an Active Migration may be suitable where a detailed schedule can be made over the migration of all functions, enabling a controlled migration towards a set deadline. In other cases, it may be suitable to allow legacy controllers to fade out as functions are migrated in the course of normal maintenance and lifecycle management of the plant. The fade out option means that Step 4 of the migration may take a very long time but it may save costs as legacy devices are used for their full lifetime, while most benefits of SOA are already available.

During this fourth step most of the functionality migrated relates to control at some level, as most of the monitoring, engineering and administration already has been moved to the SOA-system. In particular, this relates to Local control loop, Distributed control and Supervisory control. Another key function that is migrated in this step is the Emergency stop, which can be considered a form of human-in-the-loop control with some very specific conditions. As each specific control function is migrated so are the related support functions such as System aggregation, Data acquisition, display and storage and Alarms and warnings.

5.6 Conclusion

Following on and extending the initial migration concepts introduced in [3] and further detailed in [4], the novelty of migrating from a traditional hierarchical ISA95-based legacy process control system into a SOA-compliant ISA-95-based process control system is to proceed in a structured way, gradually upgrading highly integrated and vendor-locked standards into a more open structure while maintaining the functionality. Note: The migration concept presented here is not modifying the structural hierarchy of an ISA-95-based process control system but allowing it to functionally behave as a highly distributed flat architecture based on services located on physical components and/or on the cyber-space represented by a service-cloud.

A procedure for migrating the functionality of a DCS/SCADA to a cloud SOA-based implementation is proposed. The procedure comprises 4 distinct steps and make use of mediator technology. These 4 steps are designed to maintain the feeling of conformity between HMI and control execution and to ensure that the target system exhibits full transparency and supports open standards.

Table 5.1 Functional aspects mapped to Migration steps

Functional Aspect	Step 1	Step 2	Step 3	Step 4
Inter-protocol communication	*	$\bigcirc$		
User management and security	*	*	$\bigcirc$	
Operator manual override	(*)		$\bigcirc$	
Operator configuration	(*)		$\bigcirc$	
System aggregation	(*)		(*)	(*)
Data acquisition, display and storage			$\bigcirc$	*
Alarms and warnings			$\bigcirc$	*
Local control loop	*			$\bigcirc$
Emergency stop	(*)			$\bigcirc$
Supervisory control		(*)		$\bigcirc$
Distributed control				$\bigcirc$

The migration procedure is further analysed through a breakdown of the functionality of a DCS/SCADA and how the functionality can be migrated to SOA. A short description of an exemplifying proposed implementation for each functional aspect is provided and Table 5.1 provides a summary of how these functional aspects are related to each migration step. Many aspects are partially migrated (indicated by '*') or can be migrated depending on the scenario (indicated by '(*)') at different steps of the migration while there is a certain step where the main part of the functionality is migrated (indicated by '$\bigcirc$').

Using this stepwise approach, utilizing SOA and mediator technology, it is argued that the SOA approach will preserve functional integration, support grouping of devices, preserve real-time control and successfully address safety loops. With an emphasis on the DCS-part of an exemplifying legacy control system, the authors applied the approach and present the results reached.

Acknowledgments The authors would like to thank the European Commission for their support, and the partners of the EU FP7 project IMC-AESOP (http://www.imc-aesop.eu) for the fruitful discussions.

References

1. Bohn H, Bobek A, Golatowski F (2006) Sirena—service infrastructure for real-time embedded networked devices: a service oriented framework for different domains. In: International conference on networking, international conference on systems and international conference on mobile communications and learning technologies. ICN/ICONS/MCL 2006, p 43. doi:10.1109/ICNICONSMCL.2006.196
2. Colombo AW, Karnouskos S (2009) Towards the factory of the future: a service-oriented cross-layer infrastructure. In: ICT shaping the world: a scientific view. European Telecommunications Standards Institute (ETSI), Wiley, New York, pp 65–81

3. Delsing J, Eliasson J, Kyusakov R, Colombo AW, Jammes F, Nessaether J, Karnouskos S, Diedrich C (2011) A migration approach towards a SOA-based next generation process control and monitoring. In: 37th annual conference of the IEEE industrial electronics society (IECON 2011), Melbourne, Australia
4. Delsing J, Rosenqvist F, Carlsson O, Colombo AW, Bangemann T (2012) Migration of industrial process control systems into service oriented architecture. In: 38th annual conference of the IEEE industrial electronics society (IECON 2012), Montréal, Canada
5. Diedrich C, Bangemann T (2007) PROFIBUS PA instrumentation technology for the process industry. Oldenbourg Industrieverlag, GmbH. ISBN-13 978-3-8356-3125-0
6. Hérault C, Thomas G, Lalanda P (2005) Mediation and enterprise service bus: a position paper. In: Proceedings of the first international workshop on mediation in semantic web services (MEDIATE), CEUR workshop proceedings. http://www.ceur-ws.org/Vol-168/MEDIATE2005-paper5.pdf
7. IEC (2007) Enterprise-control system integration—part 3: activity models of manufacturing operations management (IEC 62264–3)
8. Jammes F, Smit H (2005) Service-oriented paradigms in industrial automation. IEEE Trans Industr Inf 1(1):62–70. doi:10.1109/TII.2005.844419
9. Karnouskos S, Colombo AW (2011) Architecting the next generation of service-based SCADA/DCS system of systems. In: 37th annual conference of the IEEE industrial electronics society (IECON 2011), Melbourne, Australia
10. Karnouskos S, Bangemann T, Diedrich C (2009) Integration of legacy devices in the future SOA-based factory. In: 13th IFAC symposium on information control problems in manufacturing (INCOM), Moscow, Russia
11. Karnouskos S, Colombo AW, Jammes F, Delsing J, Bangemann T (2010) Towards an architecture for service-oriented process monitoring and control. In: 36th annual conference of the IEEE industrial electronics society (IECON 2010), Phoenix, AZ
12. Karnouskos S, Colombo AW, Bangemann T, Manninen K, Camp R, Tilly M, Stluka P, Jammes F, Delsing J, Eliasson J (2012) A SOA-based architecture for empowering future collaborative cloud-based industrial automation. In: 38th annual conference of the IEEE industrial electronics society (IECON 2012), Montréal, Canada
13. Mahnke W, Leitner SH, Damm M (2009) OPC unified architecture. Springer, Heidelberg. ISBN 978-3-540-68899-0
14. Taisch M, Colombo AW, Karnouskos S, Cannata A (2009) SOCRADES roadmap: the future of SOA-based factory automation

Chapter 6
Next Generation of Engineering Methods and Tools for SOA-Based Large-Scale and Distributed Process Applications

Robert Harrison, C. Stuart McLeod, Giacomo Tavola, Marco Taisch, Armando W. Colombo, Stamatis Karnouskos, Marcel Tilly, Petr Stluka, François Jammes, Roberto Camp, Jerker Delsing, Jens Eliasson and J. Marco Mendes

Abstract Engineering methods and tools are seen as key for designing, testing, deploying and operating future infrastructures. They accompany critical processes from 'cradle-to-grave'. Here we provide an overview of the user and business requirements for engineering tools, including system development, modelling, visualisation, commissioning and change in an SOA engineering environment. An appraisal of existing engineering tools appropriate to IMC-AESOP, both commercial and development prototypes are presented, culminating in the presentation of tool cartography graphically, defining the impact of these tools within the enterprise and system lifecycle.

R. Harrison (✉) · C. S. McLeod
University of Warwick, Coventry, UK
e-mail: robert.harrison@warwick.ac.uk

C. S. McLeod
e-mail: c.s.mcleod@warwick.ac.uk

G. Tavola · M. Taisch
Politecnico di Milano, Milano, Italy
e-mail: giacomo.tavola@polimi.it

M. Taisch
e-mail: marco.taisch@polimi.it

A. W. Colombo
Schneider Electric, Marktheidenfeld, Germany
e-mail: armando.colombo@schneider-electric.com

A. W. Colombo
University of Applied Sciences Emden/Leer, Emden, Germany
e-mail: awcolombo@technik-emden.de

S. Karnouskos
SAP, Karlsruhe, Germany
e-mail: stamatis.karnouskos@sap.com

M. Tilly
Microsoft, Unterschleißheim, Germany
e-mail: marcel.tilly@microsoft.com

A. W. Colombo et al. (eds.), *Industrial Cloud-Based Cyber-Physical Systems*,
DOI: 10.1007/978-3-319-05624-1_6, © Springer International Publishing Switzerland 2014

6.1 Introduction

Engineering of a new generation of very large-scale SOA-based distributed systems for batch and continuous process applications requires a new generation of methods and tools. They need to support the lifecycle (e.g. design, configuration, management and support) of new SOA-based system architectures, which must provide SCADA functional capabilities to meet user requirements for much greater distributed control and decision-making abilities than current DCS system implementations.

Some 50 % of the DCS platforms running process plants today are at least 20 years old with a large number of these at the end of their lifecycle. They typically contain a mishmash of technology, which in turn requires a wide range of ad hoc engineering tools and methods to support them, limiting integration both vertically and horizontally and making reconfiguration of systems, or their optimisation, difficult to achieve effectively. There is a growing need to meet various migration requirements for both small and large system configurations in an efficient manner, and this capability, to be provided via the identified core IMC-AESOP technology, needs to be supported via specific engineering tools.

The shift from systems implemented via a predominantly centralised paradigm to control and monitoring strategies based on a system of systems engineering approach [15, 16] with thousands of dynamic SOA-compliant devices requires a new engineering methodology with requirements for new modelling and engineering approaches in order to enable such systems to be practically realised.

The approach adopted on the IMC-AESOP project was driven by user requirements derived from the studied use cases. These determined the required tools for the considered SOA process control application(s), from which a tool cartography has been created, as described in this chapter. In areas where suitable existing tools either did not exist or lacked adequate functionality, new tools were developed or existing ones extended to meet the requirements of the use cases. These are detailed

---

P. Stluka
Honeywell, Prague, Czech Republic
e-mail: petr.stluka@honeywell.com

F. Jammes
Schneider Electric, Grenoble, France
e-mail: francois2.jammes@schneider-electric.com

R. Camp
FluidHouse, Jyväskylä, Finland
e-mail: roberto.camp@fluidhouse.fi

J. Delsing · J. Eliasson
Luleå University of Technology, Luleå, Sweden
e-mail: jerker.delsing@ltu.se; jens.eliasson@ltu.se

J. M. Mendes
Schneider Electric, Marktheidenfeld, Germany
e-mail: marco.mendes@schneider-electric.com

in Sect. 6.3 along with the requirements they fulfil, how they were implemented, and the results they produced.

The IMC-AESOP project has addressed the issues of DCS across differing industrial/business domains, as highlighted by the different use cases, from real-time lubrication of mining machinery to heating management systems and district-heating applications. The requirements of these applications are broad, and the engineering content in each varies significantly. However, Sect. 6.4 describes how each of the tools is used and categorises them as per their architectural level for each use case. Finally, the IMC-AESOP toolkit is identified, highlighting the tools recommended to build the selected applications.

6.2 Engineering Methods and Cartography of Identified Tools

The first activity constitutes the identification of tools and methods that are relevant to the needs of the project, based on a study of the state of the art in engineering tools and methods for the development and support of SCADA and DCS systems. An overview is provided of the user and business requirements for engineering tools, from the SOA, system modelling and change management, application and device design development and support perspectives. System simulation, visualisation, commissioning and optimisation were also considered, together with an overview of system-of-systems engineering from a tools and methods perspective.

6.2.1 SOA Engineering Methods

The engineering of distributed embedded systems requires the modelling and support of units of distributed functionality. Object-based specifications emphasise structural decomposition, which facilitates the implementation of open and reconfigurable systems, whilst industrial software standards such as IEC 61131-3 [4] provide mechanisms for functionally decomposing and programming and IEC 61499 [6] describes the modelling of communicating distributed function blocks. In this case a system can be described as a composition of interacting components, such as function blocks or port-based objects, which are then mapped onto real-time tasks. Object and function block-based design uses a number of fundamental principles such as encapsulation, aggregation and association of objects or components to build applications.

However, whilst object- and component-based system software development is well established in several domains, a major problem that has to be overcome is the current informal and largely ad hoc definition of application components. Ad hoc specification and design may severely limit component reusability. Therefore, it is highly desirable to develop a formal framework that will allow for a systematic specification of reconfigurable components, which will be reusable by definition. Such factors needed careful consideration in the realisation of the IMC-AESOP system architecture and middleware which promote the building of applications comprising interacting components from device level up to enterprise level.

6.2.2 System Modelling, Evolution and Change Management

In order to implement a smooth and consistent transition from a SCADA environment developed via a classical (and often inhomogeneous and improvised) approach on proprietary, obsolete and rigid platforms into an open, standard, service-oriented one, appropriate tools need to be realised for system modelling, evolution and change management.

The concept of application engineering tools need to be considered in a broad context including methodologies and soft skills in addition to the technological aspects.

- From the system- and information-modelling perspectives support is needed for the following:

  - A representation of the industrial environment with a standard recognised syntax.
  - The enrichment of models with related information as parameters, semantic information, links with other objects and conditional functionalities.
  - The definition of a hierarchy (taxonomy).
  - An easy transposition in service architecture approach with automatic translation into XML syntax of metadata.

- Support for system change-management and evolution-management tools, although beyond the scope of IMC-AESOP, are crucial for effective lifecycle engineering. The management of change encompasses activities beyond data management, including people management, motivation and the ability to identify key factors in complex environments.

6.2.3 Application Design

Engineering tools to support the design of automation applications need to be implemented in order to maximise the capabilities offered by the technological solutions implemented at the shop floor level; this includes, but is not limited to, machine mechanical modular design, distributed plug-and-play control systems, interfacing to TCP-based networks and high-level process management systems (e.g. ERP and SCADA), as well as remote maintenance and monitoring systems, and support for reusability and reconfigurability for effective design-change management (Fig. 6.1).

In summary, key aspects of application engineering tool capabilities are the following:

- Reconfiguration. Configuration and change of the application by the user.
- Process description and validation. Control application defined in the user's terms.

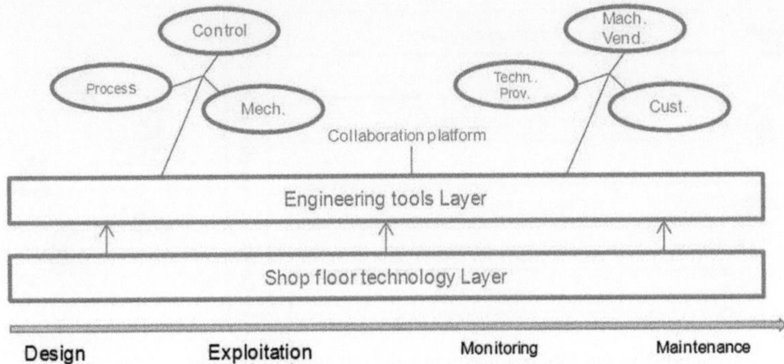

Fig. 6.1 Automation system tool requirements–lifecycle and collaboration

- Device support and maintenance. Support for embedded device.
- Real-to-virtual connectivity. Integration with virtual engineering.
- Process-to-business connectivity. Integration with business and production monitoring/support applications.

6.2.4 Device Development and Support

Using SOA on embedded devices down on the factory floor can enable powerful cross-layer possibilities. However, SOA protocols, originally developed for the enterprise domain impose heavy restrictions on their usage on resource-constrained embedded systems. This is especially true in the context of a Wireless Sensor and Actuator Network (WSAN). The relatively low bandwidth of a wireless network is a limiting factor for network performance when sensing very large packets. The commonly used protocol today for SOA in industrial automation is SOAP, using the verbose XML language. The use of XML drastically increases the size of a message containing sensor data. However, XML has excellent support today from a large number of software vendors, which makes it an open and standardised way to exchange data between devices from different manufacturers using different operating systems and applications. One benefit of the SOA approach is that message parsers can be automatically generated for each message class. This reduces the need to manually write software for the serialisation of messages.

Today, the two most widely used operating systems for wireless sensor and actuator networks are TinyOS from University of California in Berkeley, and Contiki from the Swedish Institute of Computer Science, Stockholm, Sweden. Research is currently being performed by researchers in both academia and the industry to move SOA technology down to sensor node level. The application of a widely used operating system combined with auto-generated message parsers enables system developers to reuse the existing code base to a large extent and mitigates the need to develop

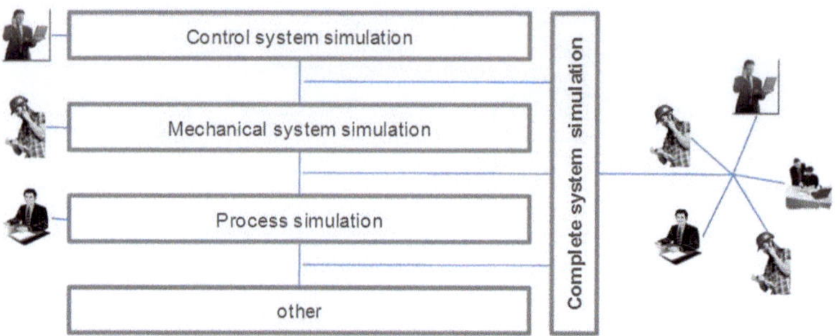

Fig. 6.2 Automation system tool requirements–design and visualisation

proprietary communication protocols. Today, gSOAP is a commonly used, proven, and stable tool to generate XML parsers. New solutions based on binary XML, i.e. EXI, need new tools to be developed. The use of EXI will be an enabling technology for SOA-based wireless sensor and actuator networks.

6.2.5 Simulation

System simulation is advantageous both at design time and during the operational phase. Simulation capabilities (i.e. the capabilities to simulate the time-dependent, dynamic system behaviours) can potentially provide strong support for testing various aspects of the system design in a virtual form prior to its final implementation in order to minimise design errors and compress design time. Simulation capabilities need to be provided for, and adapted to, each engineering application that the system design involves (e.g. control, process, and mechanical) as shown in Fig. 6.2. However, domain-dependent simulation capabilities should be integrated in a form that will enable multidisciplinary engineering teams to assess the level of completion and quality of their specific design with regard to the characteristics and behaviours expected of the final system.

6.2.6 System Visualisation

Application tools that provide system visualisation capabilities are typically user or domain- specific, i.e. they provide a representation of the system focused on the requirements of a specific end user or domain of activity. For example, a process automation system will be perceived differently by control, process, electrical or mechanical engineers, and the tools used to visualise the system need to be specifically designed to their requirements, as illustrated in Fig. 6.3.

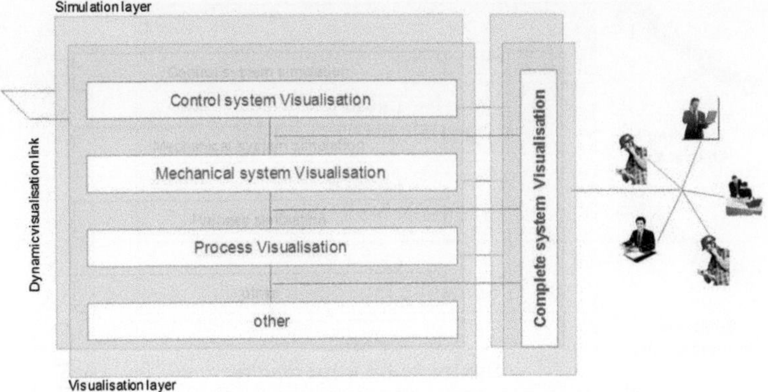

Fig. 6.3 Automation system visualisation tool

Simulation, visualisation and system engineering applications are typically tightly linked so that visualisation tools provide both static and dynamic system representations during the system design phase. Visualisation tools should provide user-specific views of a system. However, a common, intuitive, non-specific representation of the system to support cross-domains engineering collaboration and general communication between project partners is also needed.

3D computer graphic representations are typically used to implement common visualisation models of physical systems such as automation systems since a 3D virtual system representation provides a view of the system as perceived by the human physiological sensory system (i.e. sight) as opposed to the specific mental images that various engineers might have of the system. Domain-specific system visualisation tools might make use of very different representations (e.g. IEC 61131-3 for PLC control logic, state transition diagrams and Petri-net-based representations for discrete system behaviour visualisation, and flow-based chart/graphs for continuous system visualisation).

6.2.7 System Commissioning

The commissioning of a plant is a specific phase in the lifecycle. In this phase, the equipment, especially the field devices, will be installed and adapted to the real process requirements. Important tasks during commissioning include the identification of the devices, including the device type information, and setting the correct instance information, e.g. the tag names of the devices. Commissioning may be subdivided into offline and online configuration activities. The offline configuration can be started directly after the instrumentation is defined, capturing the types of the required devices, their placement inside the plant and their addresses.

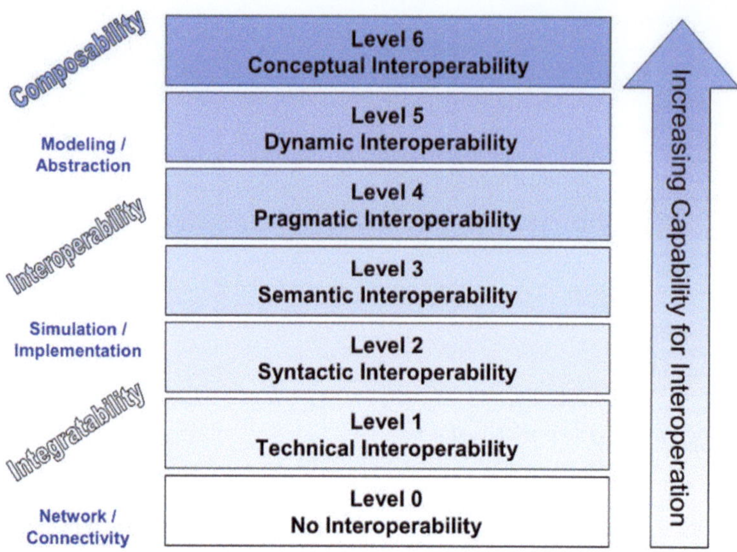

Fig. 6.4 Levels of interoperability [2]

Functional information is required for device configuration. Typical technologies providing such information, for offline as well as online configuration, are the Electronic Device Description Language (EDDL, IEC 61804-3) [8, 22], Field Device Tool (FDT, IEC 62453) [7] or the upcoming approach of Field Device Integration (FDI).

6.2.8 *Interoperability*

In order to build distributed control applications the components of a system need to communicate successfully. Engineering tools are required to support this integration, and where direct interaction is not possible, then wrappers and mediators must be employed. Figure 6.4 illustrates the levels of interoperability from no interoperability and technical interoperability at the network level through syntactic, semantic and pragmatic interoperability at the application level, to dynamic and conceptual interoperability at the modelling level. Each of these is described below [2].

- *Level* 0. Stand-alone systems have No Interoperability.
- *Level* 1. On the level of Technical Interoperability, a communication protocol exists for exchanging data between participating systems. Here, the communication of simple data is achieved with the network protocols being unambiguously defined.
- *Level* 2. The Syntactic Interoperability level introduces a common structure to exchange information, i.e. a common data format is applied. On this level, a com-

mon protocol to structure the data is used; the format of the information exchange is unambiguously defined.

- *Level* 3. If a common information exchange reference model is used, the level of Semantic Interoperability is reached. On this level, the meaning of the data is shared; the content of the information exchange requests are unambiguously defined.
- *Level* 4. Pragmatic Interoperability is reached when the interoperating systems are aware of the methods and procedures that others are employing. In other words, the use of the data or the context of its application is understood by the participating systems; the context in which the information is exchanged is unambiguously defined.
- *Level* 5. As a system operates on data over time, the state of that system will change, and this includes the assumptions and constraints that affect its data interchange. If systems have attained Dynamic Interoperability, then they are able to comprehend the state changes that occur in the assumptions and constraints that others are subject to over time, and are able to take advantage of those changes. In particular when interested in the effects of operations, this becomes increasingly important; the effect of the information exchange within the participating systems is unambiguously defined.
- *Level* 6. Finally, if the conceptual model is a meaningful abstraction of reality, the highest level of interoperability is reached: Conceptual Interoperability. This requires that conceptual models be documented based on engineering methods enabling their interpretation and evaluation by other engineers. In other words, on this we need a "fully specified but implementation independent model" and not just text describing the conceptual idea.

6.2.9 Cartography of Identified Tools

Within the IMC-AESOP project a study was carried out to compare practically available enabling technologies for application systems engineering, which might be utilised within the scope of the project. The coverage takes a selective and critical look at available candidate engineering tools.

Summaries of the strengths and weaknesses of the evaluated engineering tools provide a broad indication of their capabilities against selected criteria for control, enterprise integration, supply chain/lifecycle support, and virtual engineering. This information was helpful in understanding what aspects of current toolsets were applicable in the context of the project use cases. As a result of this study, tools suitable for use on the IMC-AESOP project were identified.

As can be seen in Fig. 6.5 above, the coverage of the tools is potentially sufficient to allow the whole lifecycle and all architectural levels to be supported. There is, however, a large amount of overlap between many of the tools, and support is highly fragmented. It should be noted that this overlap does not necessarily imply duplication, as different aspects (either complimentary or unrelated) may be covered.

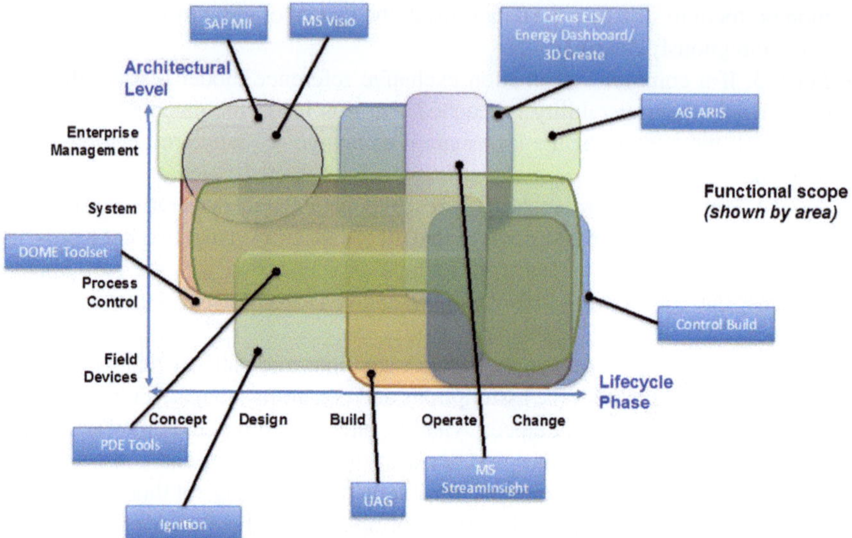

Fig. 6.5 Overview of tool cartography

Table 6.1 describes the selected engineering tools that were subsequently identified for use in IMC-AESOP.

6.3 Prototypes of Critical Tools, Including System Modelling and Analysis Methods

Based on the analysis of the requirements of the IMC-AESOP use cases against the tool capabilities described in Sect. 6.2, critical new tools or tool-extensions were identified and developed within the project. In particular:

1. EXI Compression. See Sect. 4.2.2.
2. Ignition SCADA OPC-UA API. See Sect. 4.2.4.
3. Electric car charging optimiser. See Sect. 9.3.2.
4. Orchestrators. See Sect. 9.3.2.
5. Service bus configurator. See Sect. 7.3.3.
6. Aggregations services. See Sect. 9.2.3.
7. Process Definition Environment, PDE toolkit. See Sect. 10.3.
8. Continuum. See Sect. 10.3.

The following subsections briefly review the emerging tool developments undertaken on the project highlighting the requirements addressed, the implementations undertaken, and the results achieved.

Table 6.1 Engineering tools

Name	Description
Control build application generator	Control build is dedicated to the needs of control systems engineers, providing standards-based (IEC 61131-3) programming languages and integration into HMI/SCADA systems; it is positioned well for PLC/distributed system development [11]
PDE toolkit	The PDE toolkit supports a component-based approach to systems engineering. On the IMC-AESOP project it enabled application logic for three of the use-cases to be defined in a state-based manner and supported the creation of an integrated 3D visualisation of the system behaviour in each case [11]
SAP MII	SAP Manufacturing and Intelligence (MII) [24] is a tool that provides the capability of integrating business logic within monitoring and visualising KPIs. Additionally, it is fully integrated via enterprise services with other systems such as ERP, CRM, etc. In IMC-AESOP this was used to demonstrate the creation of flexible monitoring event-driven KPIs, visualisation of business-relevant data and integration with shop floor devices
Cross-layer integration tools	Honeywell prototype engineering tools for the configuration and maintenance of plant information models to support cross-layer integration by maintaining consistency between individual layers of the process plant hierarchy [5]
ARIS—Architecture of Integrated Information Systems	ARIS is designed to provide a framework in which business components and interactions may be described and stored in detail. These components may then be used to build and analyse business processes to provide more effective business processes. ARIS was used in the IMC-AESOP project to design, store and analyse the business process including the interaction between stakeholders in the supply chain [13]
Microsoft stream-insight	Microsoft's StreamInsight [18] provides a flexible platform to enable low-latency complex event processing. These capabilities were used in IMC-AESOP to provide a general-purpose service enabling alarm processing, monitoring and system diagnostics in a very adaptable way [1]
DOME tools	The Distributed Object Model Environment (DOME) Toolset from ifak is a suite of tools used on IMC-AESOP to support the engineering and commissioning of SOA-based applications. The toolset provides translation from the object notation language (DOME-L) to a target language (currently C++), as well as tools for debugging, network discovery, examination and connection of automation devices providing DOME functionality [21]

6.3.1 EXI Compression

Requirement. In order to support device development, as has been stated in Sect. 4.2.2, EXI is a promising technology to compress the amount of data being transmitted over the network. No open-source implementation of the EXI specification had been identified that was specially targeted at resource-constrained devices. By creating a suitable tool, it was possible to encode up to ten times smaller standard service messages (XML) on extreme resource-constrained sensor and actuator devices.

Implementation Description and Use. The tool was developed from scratch using a modular design and portable source code. The tool was used extensively during the IMC-AESOP demonstrations related to the LKAB ore processing and district heating application use-cases for implementing light-weight SOAP and RESTful Web services.

Results. The developed tool is open source (exip.sourceforge.net), has been downloaded more than 1600 times, and is already being used in projects outside IMC-AESOP. The current version of the tool is in alpha form, and there is a need for more testing to make the code stable enough for production use. The tool comes with both user- and developer-documentation that is up to date and in use by contributors and end-users alike.

6.3.2 Ignition SCADA OPC-UA API

Requirement. As described in Sect. 4.2.4 OPC-UA is a widely accepted standard for providing interoperability between devices and systems. For the IMC-AESOP project Ignition SCADA was the chosen SCADA solution which supports OPC-UA. To integrate Ignition SCADA with DPWS an OPC-UA API was to be created. Ignition is an industrial application server from Inductive Automation [9], used to create HMI, SCADA and MES systems. Ignition (formerly FactorySQL and FactoryPMI) is a mature and well-tested application. The feature list includes:

- Web-based gateway configuration and HMI drag-and-drop editor.
- A rich set of visual components free for end-users.
- Database-centric architecture.
- Advanced reporting and alerting mechanisms.
- Designed from the ground-up for scalability.
- Implemented in Java, so it can be run on a wide range of platforms, on all major operating systems.
- Control system access on mobile device.

State-of-the-art SCADA systems are capable of supporting almost any industrial communication protocol, such as Modbus, Profibus and DeviceNet. In the case of DPWS embedded devices, however, such support does not currently exist. It was therefore necessary to develop a SCADA plug-in that would enable this functionality in a commercial SCADA system, in this case the Inductive Automation Ignition system.

Implementation Description and Use. A new gateway module was created on IMC-AESOP containing a DPWS client stack. This client stack will discover DPWS devices and create representations of the devices as nodes in the OPC-UA

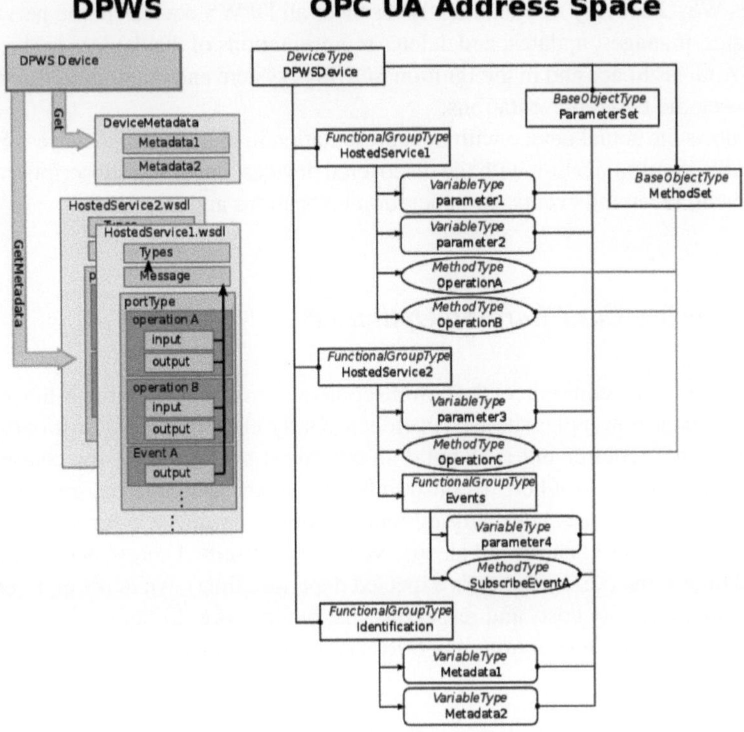

Fig. 6.6 Generic mapping from DPWS to OPC-UA for device object model [19]

Address Space. Any OPC-UA client connecting to the Ignition OPC-UA server must be able to invoke operations and subscribe to events on the discovered DPWS devices, using the OPC-UA service sets. The Ignition SDK, a collection of libraries and sample code for creating custom Ignition Modules, was used to achieve this mapping. A 'generic mapping' is shown in Fig. 6.6 (https://www.inductiveautomation.com/scada-software). The MethodSet gathers all the methods that are exposed to the client, and the ParameterSet gathers all parameters of the device. The FunctionalGroups representing the hosted services organises the methods and parameters of the device. Multiple FunctionalGroups can refer the same methods and parameters. Asynchronous push-mode events defined in WS-Eventing do not clearly fit into the OPC-UA for Devices Object Model. One approach is shown in Fig. 6.6, although many different approaches could be designed. Events are grouped in a separate functional group, nested within the Hosted Service, with the appropriate output parameters and a method for subscribing and unsubscribing to each event.

Results. A new DPWS Driver was written using the Ignition SDK, and includes JMEDS. The DPWS Driver module:

- Uses WS-Discovery to dynamically discover all DPWS devices in the network.
- Creates, manages, updates, and deletes representations of the devices in the OPC-UA Address Space and in the Ignition SQLTags system and maintains consistency between the two representations.
- Connects the actual device with its representation in the OPC-UA Address Space.
- Handles communication with the discovered devices, including subscription management, receiving events, and operation invocations and responses.

6.3.3 Electric Car Charging Optimiser

Requirement. The requirement for optimisation of electric car charging is that excess energy from the power plant be used in order to charge electric cars in a cost-optimised way. Here, the requirement is to build an optimised scheduling of the charging of electric cars in order to adhere to constraints set (e.g., the available energy, electricity price, minimum electric car charging requirements). To do so, a service has been developed that tries to charge all electric vehicles to specified energy levels within a limited timeframe (i.e. by the car's expected departure time), while trying to exploit fluctuating electricity costs and respecting maximum power limits.

Implementation Description and Use. The orchestrator brings together the following systems:

1. Plant simulator.
2. Electric Car Optimizer (running in the SAP HANA Cloud).
3. Energy Market (running as an Internet public service).

The service has been implemented in Java and runs in the SAP HANA Cloud [23]. It is called by the orchestrator, which transmits data about the available cars, their requirements and their needs, as well as information about the power plant's energy production limits and costs. The interface is implemented as a set of REST services.

Results. The electric car charging optimiser could play a crucial role in the smoothing out of power consumption and making energy production more efficient. It was shown that a cloud service providing this functionality is viable and can empower more sophisticated scenarios. Additional info can be found in Chap. 9.

6.3.4 Orchestrators

Requirement. An application (Matlab/Simulink) has been used in order to simulate the power plant itself as well as the cars that get charged. As this is currently a private simulation and not a public service, an 'orchestrator' is required to act as the glue between the three different systems. Further to this, the Orchestrator also hosts the logic for contacting the energy bidding agent and passing it all the necessary data for acting in the energy market.

Implementation Description and Use. The orchestrator logically integrates the Plant Simulator with the Electric Car Optimizer running in the SAP HANA Cloud, and the Energy Market that runs as an Internet public service [10].

Results. The implemented orchestrator plays a crucial role:

1. In acting as a mediator between the different technologies, e.g. between OPC-UA and REST.
2. As a management point during the whole lifecycle of the simulated scenario.
3. As an instrument to collect real-time data for date collection and analysis.

Additional info can be found in Chap. 9.

6.3.5 Service Bus Configurator

Requirement. The Distributed Service Bus middleware applied on IMC-AESOP required some configuration and monitoring capability. The identified needs included:

1. Monitoring of discovered Service Bus nodes and managed devices.
2. Enabling/disabling specific interfaces.
3. Security configuration (Role Based Access Control).
4. Management of Service Bus built-in services, including time synchronisation.

In order to facilitate the deployment and maintenance of this tool, a Web client solution, served by one or several Service Bus bricks, was preferred.

Implementation Description and Use. The Configurator has been implemented as a Web application, using the popular Google Web Toolkit development framework. This framework allows developers to write the client application in the Java language and then convert it to AJAX (Asynchronous JavaScript).

Results. The implementation proved to be successful, providing a stable and efficient tool to monitor the Distributed Service Bus, and it has been demonstrated in the lubrication application of use Case 1. The Configurator tool was used, in particular, to manage time synchronisation between the Service Bus and other demonstrator components (Mediator and CoAP), monitor managed devices (AS-I and CoAP), and monitor the LKAB application logic that was running on the Service Bus itself.

6.3.6 Aggregation Services

Requirement. The orchestration of services is a core concept in Service-Oriented Architectures (SOA). This is basically about creating new, added-value services out of existing services and is a concept that was established from the beginning of SOA. There are a number of specifications available to assist in the composition of services. The most prominent one is probably WS-BPEL [20]. Others either deal with the orchestration or choreography of services. Within IMC-AESOP the focus

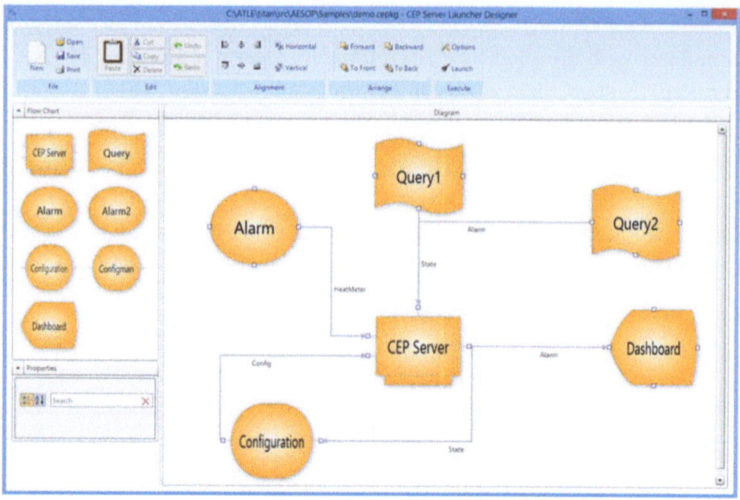

Fig. 6.7 CEP server launch designer

was on providing a service or a tool for helping in the real-time analysis of a large number of events from multiple sources. In this case, the service that processes data is a Complex Event Processing (CEP) service. Currently, services are being deployed at device level, such in the Internet of Things, which is advocated by the IMC-AESOP project. These services may be considered as data services providing a continuous stream of data. The CEP service may be regarded as an event broker with analytics capabilities and with the ability to connect stack services, making it possible to:

1. Define the flow of events (data) by topics.
2. Define queries (analytics) processing incoming data by topic.
3. Define consumers of events by topics.

Implementation Description and Use. To enable the definition of events and topics in an easy way a 'CEP Server Launcher Designer' tool has been developed. This tool (see Fig. 6.7) helps to define a list of services that act as sender (e.g. Alarm), consumer (e.g. Dashboard) or broker (e.g. Configuration). The CEP Server is a special instance since it accepts query definitions. Each connection is annotated with the name of the topic. The direction of the arrow (from Alarm to CEP Server) shows the direction of the event flow.

Results. So far the tool has been used internally. Functionally it provides an effective solution, but quantitative results have not yet been collated.

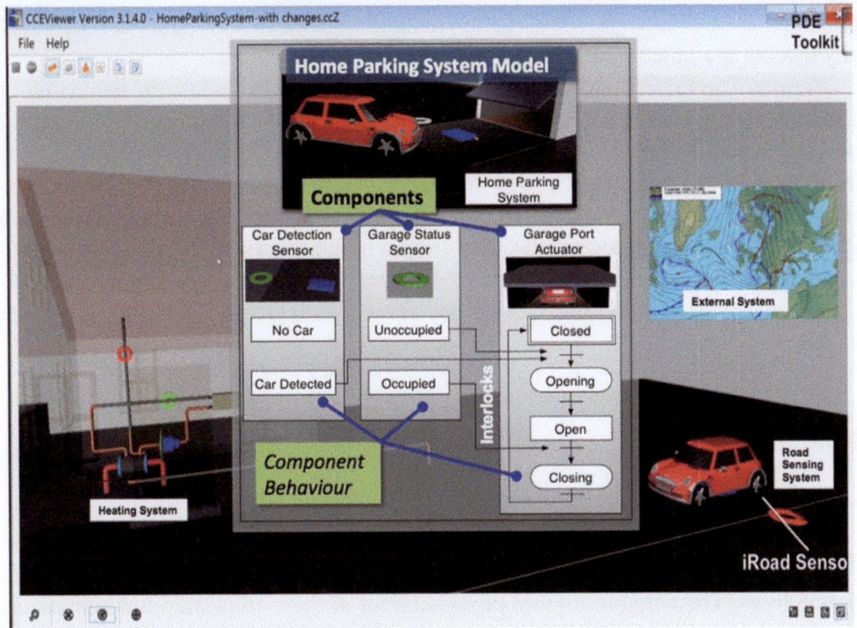

Fig. 6.8   Process definition environment (PDE)

6.3.7 Process Definition Environment, PDE Toolkit

Requirement. The IMC-AESOP project requires the building and integration of application-related services across a range of systems. The Process Definition Environment (PDE) enables the definition of relationships between system components in support of orchestration or choreography. The PDE toolkit enables systems composed of many logically interlocked components to be visualised and validated.

Implementation Description and Use. On the IMC-AESOP project the PDE toolkit was used to validate and visualise Use Cases 1 and 4 for lubrication and building systems, respectively [11]. This is illustrated in Fig. 6.8. Routing logic defined within the PDE toolset enables the material or product flow in the system to be simulated. This in conjunction with a simulation engine enables each application to be observed and validated. The state-based components within the system can be stored in a library with their associated 3D visualisations. The PDE toolkit [12] employs a simulation engine that requires the definition of how the outside world (e.g., product or process parts) interacts with the machinery. This process knowledge, typically defined by a process engineer, is entered into a routing, which in turn drives the input of the sensors in the simulation, which drives the application to react to fulfil the desired logic. Once the control has been validated, it can potentially be downloaded to orchestration engines on a range of SOA devices; however, this phase was beyond the scope of IMC-AESOP although it is a major objective for future work. The PDE tool supports the deployment of the application logic to orchestrators at potentially any level in the target architecture on a range of target devices.

Results. The PDE toolkit has been successfully used to simulate the behaviour of components in the lubrication application of Use Case 1 and verify the behaviour of the district heating system in Use Case 4 as depicted in Chap. 10. Break points have allowed the scenarios defined in these use cases to be quickly evaluated. The inclusion of processes has created the ability to 'manage' the automation system using the high-level process view. This results in the control system being easier to understand from a maintenance point of view and in the case of Use Case 4 provides the ability to run different processes based upon external factors, such as long-range weather forecasts.

6.3.8 Continuum

Requirement. With the Continuum tool [17] it was possible to create very large control and monitoring structures by using the formal method 'High-level Petri Nets' (HLPN). Related to the HLPN theory there are a range of analysis and validation possibilities. The Continuum tool provided a lot of these important functionalities. Based on analysis and validation results, calculated from the tool, it was much easier to introduce the simulation of complex monitoring and control systems in the next step. The start of the implementation of the Continuum tool was in the EU FP6 SOCRADES project where the first control structures were built with this tool. In this project the Continuum tool helped to solve the challenge to design very complex control structures for very large control and monitoring systems in a rapid manner.

Implementation Description and Use. The performance challenge was solved by parallelising the algorithms and by using GPUs for the new algorithm design. The company NVIDIA provides a framework called CUDA for parallel programming needs, which was used for the parallelisation. The performance of the Continuum tool is now growing with the increasing performance of NVIDIA GPU technologies.

Results. With support from this tool it was possible to generate monitoring and control structures in a very fast way for very large systems. Continuum is a powerful prototype tool with the potential to be extended in the future to a mature engineering tool. First publications are available describing a graphical engineering interface and method that will provide a very easy engineering method. The approach is simple to apply even to very complex control and monitoring structures in a system-of-systems paradigm, with automatic support for analysis and validation by the tool.

6.4 Engineering Toolkit Application to Use Cases

In order to explain and classify methods and tools employed in the different engineering processes associated with the design, development and commissioning of SOA-based monitoring and control systems on the IMC-AESOP project, architec-

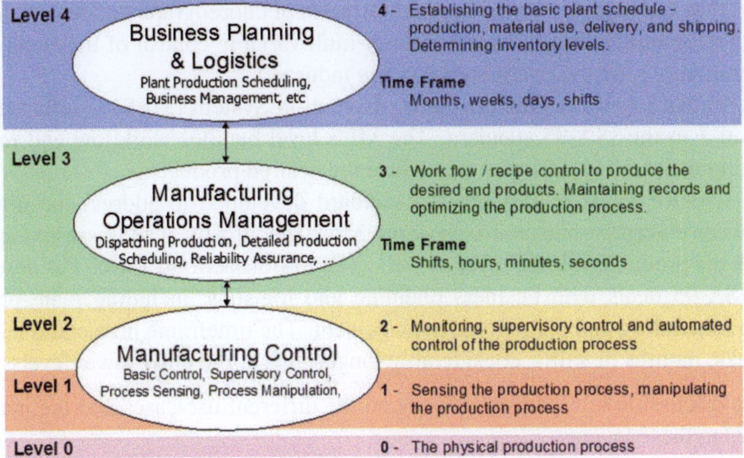

Fig. 6.9 ISA-95 layers [3]

tural levels (e.g. field device, process control system and enterprise management), as defined by the standard ISA-95, have been used. The four use cases introduced in Chap. 1 are presented below, with regard to their functional and architectural aspects, in relation to engineering methods and tools utilised to engineer each of them. The set of tools used in each use case is presented in a diagram, showing the mapping between tools and the addressed ISA-95 architectural levels.

6.4.1 ISA-95 Layers Applied to the Categorisation of IMC-AESOP Tools

A brief description of the ISA-95 standard is included here for completeness. The standard consists of several layers [3] as illustrated in Fig. 6.9:

- *ISA-95 Device Level* 0–1. ISA-95 device level consists of levels 1 and 0 in the standard. Level 1 is the level for sensing and manipulating the production process, usually consisting of sensors and actuators. Level 1 is connected to Level 0 which is the actual production process, more specifically described as the actual physical process. The device level consists of usually small resource-constrained devices that link the service architecture to the production process. The importance of a semantic Web service approach at the device level can be found in [14].
- *ISA-95 Control/ SCADA Level* 2. In general terms, this level is concerned with the control and visibility of production processes. This does not include the real-time control of processing equipment, which is the concern of Level 1, but chiefly the integration, e.g., the orchestration or choreography of devices, in the Level 1 controllers to achieve specific tasks related to recipes or production objectives.

This may be in the form of either orchestration or choreography of devices (e.g. in discrete manufacturing), or supervisory multivariable control of the steady state operation (e.g. in the continuous process industry).

- *ISA-95 MES Level* 3. Manufacturing Execution System (MES) is referred to as Level 3 in the ISA-95 standard. The MES level handles workflow and recipes, among other things, in order to produce the desired products.
- *ISA-95 ERP Level* 4. ISA-95 is a standard developed to address the interface between enterprise and control systems. More specifically, it addresses integration with the production/MES layer as well as B2M transactions. Level 4 as described in ISA-95 deals with business planning and logistics, including plant production scheduling and operational management. The timeframe here could be days, weeks, months or shifts, considerably longer than at the other lower levels.

The discussion now moves to focus on the different use-cases and the mapping of tools to these layers.

6.4.2 Use Case 1: Migration of a Legacy Plant Lubrication System to SOA

The objective of use Case 1 was to demonstrate the migration of an existing lubrication system in an industrial process plant to the IMC-AESOP architecture. Lubrication systems represent one of the most important types of support systems seen in such industries. Although such lubrication systems are critical for good performance, they are often implemented as black-boxes with limited system integration.

The tools utilised to engineer the migration of the plant lubrication system to SOA are described below.

- *C compiler (GCC).* A compiler is required to translate the human readable source code into binary code for processors and microcontrollers. The Mulle module uses a Renesas M16C microcontroller. The GNU Compiler Collection consists of a C and several other compilers. GCC supports many different development and target systems. With cross compiling it is possible to develop software on an ordinary PC and compile it to a completely different system e.g. running on an $\times 86$ PC.
- *Timber.* Timber is a functional programming language derived from Haskell. Timber is event-driven using reactive objects. The compiler compiles from Timber to intermediate C code that is parsed to GCC. Other intermediate options like Low Level Virtual Machine (LLVM) are investigated by other organisations. Earlier versions of Timber were interpreted.
- *Flasher.* Flasher is a software kit to program flash memories on the Mulle module. Mulle uses a Renesas M16C microcontroller connected to external flash memories.
- *CoAP/EXI.* CoAP can support several transfer methods, ranging from human readable XML to more efficient binary methods. In order to enable efficient XML Interchange, EXI has been chosen for this use case due to several performance advantages, such as reduced memory footprint and shorter packets that are more

likely to fit in a single (radio) frame. For embedded devices and sensors in particular memory resources are often scarce. Fewer radio frames increase reliability and battery life.

- *SOA4D DPWS Toolkit*. One of the major results of recent R&D projects like ITEA2 SIRENA and European FP6 SOCRADES has been a DPWS implementation called SOA for Devices (http://www.soa4d.org). This implementation has been further improved for the IMC-AESOP approach and used in this use case. Specifically, the SOA4D toolkit is utilised to implement the DPWS support in the service bus, both for the communication with a mediator component and for the communication with a Smart Meter Emulator.
- *Smart Meter Emulator*. A simulator for the energy-related aspects of devices targeting smart metering and costs has been developed. It gives the possibility to interact via DPWS and REST with the devices and additionally supports device lifecycle management (e.g. for start/stop/add/remove) of any device, flexible description of classes of devices and their behaviour (done in XML configuration), adjustment of energy prices that are used for operational cost calculation and support for automated creation of a large number of devices to ease tests (done via XML configuration).
- *TinyOS*. TinyOS is designed for low-power wireless devices like Mulle. TinyOS partially supports IPv6 and CoAP. UDP is supported with TCP in prototype form. For CoAP GET & PUT methods are supported, but neither POST nor DELETE are supported.
- *Mozilla Copper plug-in*. This plug-in provides a handler for the 'CoAP' URI scheme to Mozilla Firefox. With this plug-in capability Firefox can be used for troubleshooting and for some configuration during commissioning.
- *Service Bus Configurator*. The Service Bus middleware is intended to run on industrial-embedded devices. Such devices typically have very limited HMI capabilities. The Service Bus, however, needs some means for its configuration and monitoring. This has been achieved through a Web-based application hosted on each device, called a Service Bus Configurator. This application relies on Rich InteLTUrnet Application concepts and has been written using the Google Web Toolkit software development kit. The Configurator application is downloaded from any device hosting the Service Bus into a Web browser running on a computer connected to this device. Communication between the Configurator and the Service Bus is based on RESTful Web services. Features that can be configured/monitored using the Service Bus Configurator include:

  - Adding a new device to the Service Bus (e.g. CoAP edge router and AS-i gateway…).
  - Monitoring devices/services composing the Service Bus (status, configuration parameters…).
  - Configuring/monitoring event broker topics and subscriptions.
  - Service Bus system events.

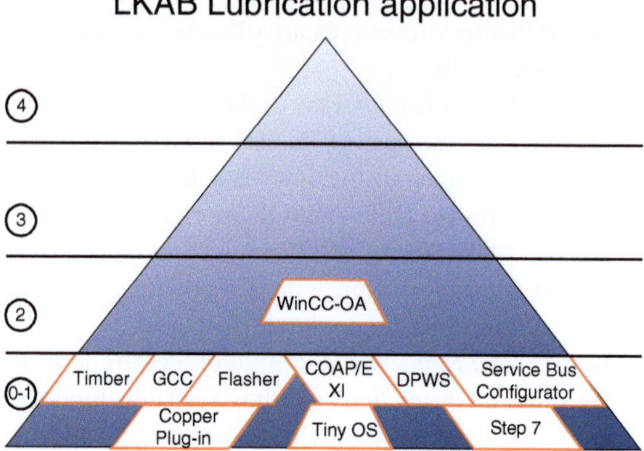

Fig. 6.10 Use Case 1 plant lubrication application: tool mapping into the ISA-95-levels

- *WinCC Open Architecture*. WinCC-OA [25] is a complex and flexible object-oriented SCADA system that is implemented on the Maintenance station as an HMI solution for the lubrication system.
- *Step 7*. The SIEMENS Step 7 [26] programming environment has been used to investigate the legacy PLC-code in the lubrication system. The result of this analysis was a humanly readable functional description that was then used to implement the SOA-based replacement system functionality that resulted from the migration process.
- *PDE Toolkit*. Provided system application logic simulation, validation and process visualisation. The lubrication system configuration was modelled from its constituent components, e.g. values and pumps, supporting both their state behaviour and 3D schematic visualisation.

Figure 6.10 illustrates how the tools used in Use Case 1 correspond to the ISA-95 levels.

6.4.3 Use Case 2: Implementing Circulating Oil Lubrication Systems Based on the IMC-AESOP Architecture

Oil lubrication systems are commonly used in paper machines where hundreds of lubrication points are needed. In general, for the implementation of this use case two different types of tools were used:

1. Development and Deployment tools were used to create, program and configure the necessary behaviours and functions of the use case.
2. Testing tools used to test the developed and deployed system.
   Development tools used:

Fig. 6.11 Use Case 2 circulating oil lubrication application: tool mapping into the ISA-95-levels

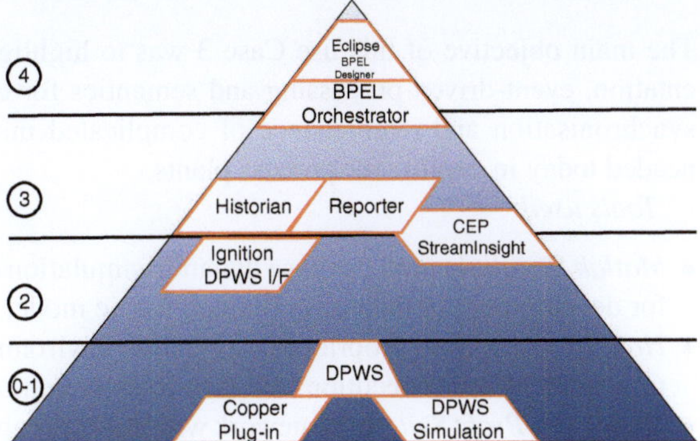

- *Apache ServiceMix.* A platform providing useful functionality for integrating technologies internal to components, and WS frameworks for exposing services.
- *Apache Camel.* An open source integration framework based on enterprise integration patterns, which provides connectivity to a wide array of technologies /transports/protocols/APIs, included in ServiceMix.
- *Jetty.* A simple HTTP server, which can be used for consuming and producing HTTP requests.
- *WS4D-JMEDS* (*Web services for Devices—Java Multi-Edition DPWS Stack*). An open-source stack for developing DPWS clients, devices, and services.
- *JAX-WS.* (Java API for XML Web services) A Java API for developing Web services.
- *WCF* (*Windows Communication Foundation*). Windows runtime API for developing SOA applications in C#.
- *StreamInsight.* A platform for developing and deploying CEP applications from Microsoft.
- *Ignition Server.* A commercial HMI/SCADA system with integrated OPC-UA server.
- *Ignition Developer API.* An application programming interface for developing custom modules for the Ignition Gateway, Designer, or Client in Eclipse.
- *Eclipse BPEL Designer.* A graphical editor for creating BPEL processes.
- *Orchestration Engine.* A tool developed for executing WS-BPEL processes.

  Testing tools used:

- *WCFStormLite.* For testing WCF Services.
- *DPWS Explorer.* For testing services on DPWS devices.
- *UA Expert.* A free OPC-UA Client for testing OPC-UA and DPWS integration.

  Figure 6.11 shows the ISA-95 level for these tools in a graphical form.

6.4.4 Use Case 3: Plant Energy Management

The main objective of this use Case 3 was to highlight advantages of service orientation, event-driven processing and semantics for easier configuration, dynamic synchronisation and maintenance of complicated multilayer solutions, which are needed today in continuous process plants.

Tools used:

- *Matlab Simulink*: multipurpose dynamic simulation environment, which was used for development of the power plant dynamic model.
- *Honeywell UniSim*: proprietary simulation environment for development, validation and real-time execution of dynamic process models.
- *Honeywell Profit Suite*: a framework with a set of proprietary tools for development, configuration, and deployment of control applications.
- *Microsoft StreamInsight*: a framework for implementation of Complex Event Processing applications.
- *Eclipse*: for JAVA implementation and Microsoft *Visual Studio for C++* implementation.
- *Information model building tools*: Address Space Model Designer (ASMD), XML editor and OPC-UA Model Compiler. tools were used to create the OPC-UA address space. This includes nodes, attributes and their mutual relationships.
- *The data binding tool*: for binding of the data items inside a server address space to external data sources.
- *Information model configuration tool*: for the chained Level 2 servers where it allows to create an instance of a subsystem and define device, topology and binding views.
- *Electric Vehicle Scheduler/Optimiser*: schedules in a optimal way (under constraints) the electric cars of the company. It is implemented as a cloud service (REST) based on SAP HANA Cloud.
- *Orchestrator*: used to integrate among the Matlab simulator (via OPC-UA), the Electric Vehicle Scheduler/Optimiser (via REST service calls), and the Energy Market (via REST service calls).
- *Energy Market*: Offers the ability to trade (buy and/or sell) energy on local energy markets as envisioned in the SmartGrid era. It is running as an Internet REST service.

Figure 6.12 illustrates how the tools used to engineering the use Case 3 correspond to individual ISA-95 levels.

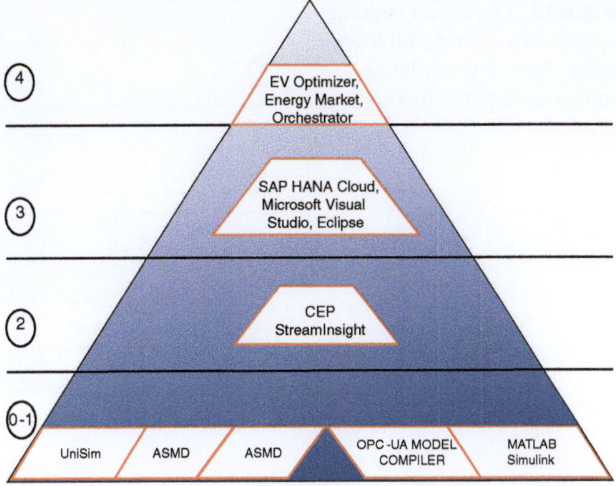

Fig. 6.12 Use Case 3 plant energy management: tool mapping into the ISA-95-levels

6.4.5 Use Case 4: Building a System of Systems with SOA Technology—a Smart House Use Case

The goal of the District Monitoring use case is to demonstrate the application of the IMC-AESOP architecture in a complex and heterogeneous environment that can enhance the overall comfort and reduce operating costs in residential areas. The application of the IMC-AESOP architecture will allow integration and configuration of operational parameters of various district systems, such as heating, electricity and transportation.

Below is a description of the tools used in the development and implementation of the District Monitoring use case. The tools are categorised in three types: Development and Deployment tools used in the development phase, Core tools that implement the required functionality during the exploitation phase and Testing tools used to verify the behaviour of the system.

Tools used:

- *Integrated Development Environment (IDE).* the prototype development used IDEs, such as Eclipse, to assist in the software implementation. This tool was used in all the components of the system in the Development and deployment phase.
- *Mulle developing kit.* This includes a C cross-compiler for Renesas M16C microcontroller (m32c-elf-gcc), a flasher (sflash) to deploy the programs and help libraries available from EISTEC AB. This tool was used for the Mulle components, iRoad/iPark devices and the Car HMI. The C cross-compiler and the flasher are Development and Deployment tools while the help libraries are Core tools.
- *Contiki.* A light-weight OS for IoT devices. This OS provides concur-rent programing and network stack for resource constrained devices. It is a Core tool

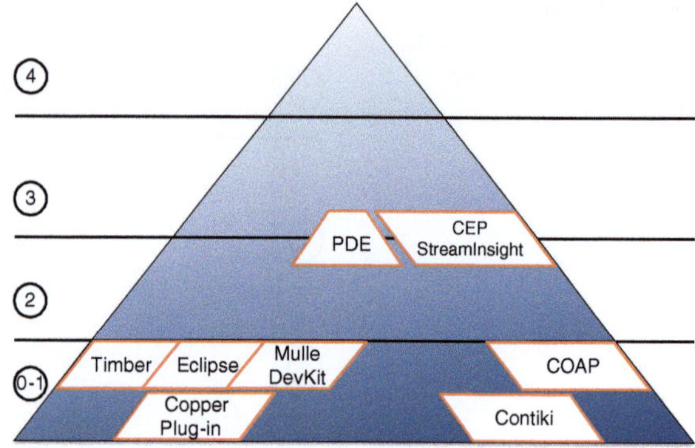

Fig. 6.13 Use Case 4 building system of systems with SOA technology: a smart house—tool mapping into the ISA-95-levels

that is the foundation of the functionality provided by the Mulle and iRoad/iPark components.

- *Timber compiler and run-time.* Used for the real-time CEP simulations in the District management system. It is both Development and Deployment tool and Core tool that execute during the exploitation phase.
- *CoAP/EXI RESTful engine.* An integration of libCoAP or Contiki built-in CoAP implementations and EXIP. It is a Core tool used to provide RESTful Web services in Mulle components, Visualisation console, iRoad/iPark and Car HMI devices.
- *Mozilla Copper plug-in.* It is a web-based interface that is used to test the CoAP interfaces in Mulle components, Visualisation console, iRoad/iPark and Car HMI devices. The tool's category is Testing.
- *Complex Event Processing Engine.* The CEP engine is based on the Microsoft StreamInsight software for developing CEP applications. This tool is categorised as both a Core and a Development and Deployment tool.
- *PDE tookit and simulation engine.* This component-based engineering environment was used to simulate and visualise the control behaviour of the system. The PDE toolset includes an integrated 3D system visualisation capability and a simulation engine. Once verified the control logic can be deployed to an associated orchestration engine on runtime control systems on a range of platforms.
- *Continuum.* This tool was used in conjunction with the output of the PDE tool to formally analyse different orchestration topologies and verify different system behaviours under varying conditions. Among others, structural and behavioural orchestration specifications can then be validated, e.g. the cyclic and deadlock-free evolution of the system.

Figure 6.13 depicts the tools used in use Case 4 and how they correspond to individual ISA-95 levels.

Fig. 6.14 IMC-AESOP tool mapping into the ISA-95 levels

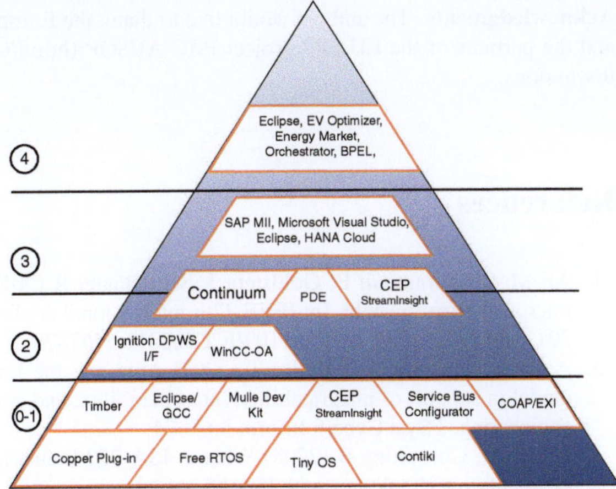

6.5 Conclusions

The methods, tools and practices needed to engineer the next generation SCADA/DCS systems will necessarily vary with the characteristics of the plant and the use cases involved. Nevertheless, on the IMC-AESOP project it has been possible, across the four use cases studied, to extract the common elements to produce an effective IMC-AESOP engineering toolkit, to aid anyone wishing to engineer a SOA-based ISA-95 multilayered system solution. Figure 6.14 summarises the IMC-AESOP tools that could be used to build SOA-based SCADA/DCS applications.

Device level support can be provided by CoAP and EXI, using the service bus configurator for integration of devices built using development tools such as Timber and the Mulle development kit, which are deployed on devices running FreeRTOS, TinyOS and Contiki.

SCADA functionality can be provided by Ignition and WinCC-OA using the prototype OPC-UA interface for integration with the devices, whilst system modelling, simulation and visualisation is supported by the PDE toolkit for simulation and Continuum for verification. The HANA Cloud is used to provide manufacturing execution system functionality.

The enterprise level tools highlighted by IMC-AESOP are the Eclipse BPEL designer and a BPEL orchestrator to execute them. Finally, Microsoft's StreamInsight has been employed for the aggregation and processing of events generated by large-scale systems as well as for integration of disparate systems, as part of a system-of-systems approach.

The combined use of these tools allows the engineering of complete applications in which components from any level of the ISA-95 can be integrated to provide a coherent SOA-based SCADA/DCS solution, such as the ones described in Chaps. 7, 8, 9, 10.

Acknowledgments The authors would like to thank the European Commission for their support, and the partners of the EU FP7 project IMC-AESOP (http://www.imc-aesop.eu) for the fruitful discussions.

References

1. Ali M, Chandramouli B, Goldstein J, Schindlauer R (2011) The extensibility framework in microsoft streaminsight. In: IEEE 27th international conference on data engineering (ICDE), 2011, pp 1242–1253. doi:10.1109/ICDE.2011.5767878

2. Andreas T, Saikou D, Charles T (2007) Applying the levels of conceptual interoperability model in support of integratability, interoperability, and composability for system-of-systems engineering. J Syst Cybern Inform 5:65–74

3. Brandl D, Consulting B (2008) What is ISA-95? Industrial best practices of manufacturing information technologies with ISA-95 models

4. Chawla R, Banerjee A (2001) A virtual environment for simulating manufacturing operations in 3D. In: Winter simulation conference, pp 991–997. doi:10.1145/564124.564265, http://dblp.uni-trier.de/db/conf/wsc/wsc2001.html%23ChawlaB01

5. CommSvr (2013) Address space model designer. http://www.commsvr.com/Products/OPCUA/UAModelDesigner.aspx

6. Dai W, Vyatkin V (2010) Redesign distributed IEC 61131–3 PLC system in IEC 61499 function blocks. In: IEEE conference on emerging technologies and factory automation (ETFA), 2010, pp 1–8. doi:10.1109/ETFA.2010.5641239

7. IEC (2009) Function field device tool (FDT) interface specification, IEC 62453. Technical report, Geneve

8. IEC (2010) Function blocks (FB) for process control—part 3: electronic device description language (EDDL), IEC 61084-3 Ed. 2.0. Technical report, Geneve

9. Ignition (2013) Inductive automation, ignition server. http://www.inductiveautomation.com/scada-software

10. Karnouskos S, Goncalves Da Silva P, Ilic D (2012) Energy services for the smart grid city. In: 6th IEEE international conference on digital ecosystem technologies–complex environment engineering (IEEE DEST-CEE), Campione d'Italia, Italy

11. Kaur N, Harrison R, West A, Phaithoonbuathong P (2010) Web services-based control devices for future generation distributed automation systems. In: Proceedings of the world congress on engineering (WCE), vol 3. London, UK

12. Kaur N, McLeod C, Jain A, Harrison R, Ahmad B, Colombo A, Delsing J (2013) Design and simulation of a soa-based system of systems for automation in the residential sector. In: IEEE international conference on industrial technology (ICIT 2013), pp 1976–1981. doi:10.1109/ICIT.2013.6505981

13. Li C, Qi J, Shu H (2008) A SOA-based ARIS model for BPR. In: IEEE international conference on e-business engineering, 2008. ICEBE '08, pp 590–595. doi:10.1109/ICEBE.2008.14

14. Lobov A, Lopez FU, Herrera VV, Puttonen J, Lastra J (2009) Semantic web services framework for manufacturing industries. In: 2008 IEEE international conference on robotics and biomimetics, IEEE, pp 2104–2108

15. Maier M (2005) Research challenges for systems-of-systems. In: IEEE international conference on systems, man and cybernetics, 2005, vol 4, pp 3149–3154. doi:10.1109/ICSMC.2005.1571630

16. Maier MW (1998) Architecting principles for systems-of-systems. Syst Eng 1(4):267–284

17. Mendes J, Bepperling A, Pinto J, Leitao P, Restivo F, Colombo A (2009) Software methodologies for the engineering of service-oriented industrial automation: the continuum project. In: IEEE 33rd annual international computer software and applications conference, 2009. COMPSAC '09, vol 1, pp 452–459. doi:10.1109/COMPSAC.2009.66

18. Microsoft (2013) Microsoft streaminsight. http://technet.microsoft.com/en-us/library/ee362541.aspx
19. Minor J (2011) Bridging OPC-UA and DPWS for industrial SOA. Master's thesis, Tampere University of Technology, Tampere. http://dspace.cc.tut.fi/dpub/bitstream/handle/123456789/20954/minor.pdf
20. OASIS (2013) OASIS web services business process execution language (WSBPEL) TC. https://www.oasis-open.org/committees/tc_home.php?wg_abbrev=wsbpel
21. Riedl M (2005) Distributed object model environment. Ph.D. thesis, Otto-von-Guericke-Universität Magdeburg, Magdeburg
22. Riedl M, Naumann F (2011) EDDL: electronic device description language. Oldenbourg Industrie, verlag
23. SAP (2013a) SAP HANA cloud. http://www.sap.com/pc/tech/in-memory-computing-hana/software/overview/index.html
24. SAP (2013b) SAP Manufacturing, Integration and Intelligence (MII). http://www.sap.com/solution/lob/scm/software/manufacturing-integration-intelligence/index.html
25. SIEMENS (2013a) SCADA System SIMATIC WinCC open architecture. http://www.automation.siemens.com/mcms/human-machine-interface/en/visualization-software/simatic-wincc-open-architecture/
26. SIEMENS (2013b) SIEMENS SIMATIC STEP 7. http://www.automation.siemens.com/mcms/simatic-controller-software/en/step7/

Chapter 7
Migration of a Legacy Plant Lubrication System to SOA

Philippe Nappey, Charbel El Kaed, Armando W. Colombo, Jens Eliasson, Andrey Kruglyak, Rumen Kyusakov, Christian Hübner, Thomas Bangemann and Oscar Carlsson

Abstract IMC-AESOP investigations have been articulated around key use cases in order to better capture user needs and corresponding requirements. This particular use case explores how Service-Oriented Architecture (SOA) can ease the installation and maintenance of one of the lubrication system of the world's largest underground

P. Nappey (✉) · C. El Kaed
Schneider Electric, Grenoble, France
e-mail: philippe.nappey@schneider-electric.com

C. El Kaed
e-mail: charbel.el-kaed@schneider-electric.com

A. W. Colombo
Schneider Electric, Marktheidenfeld, Germany
e-mail: armando.colombo@schneider-electric.com

A. W. Colombo
University of Applied Sciences Emden/Leer, Emden, Germany
e-mail: awcolombo@technik-emden.de

J. Eliasson · A. Kruglyak · R. Kyusakov
Luleå University of Technology, Luleå, Sweden
e-mail: jens.eliasson@ltu.se

A. Kruglyak
e-mail: andrey.kruglyak@ltu.se

R. Kyusakov
e-mail: rumen.kyusakov@ltu.se

C. Hübner · T. Bangemann
ifak, Magdeburg, Germany
e-mail: christian.huebner@ifak.eu

T. Bangemann
e-mail: thomas.bangemann@ifak.eu

O. Carlsson
Midroc Electro AB, Stockholm, Sweden
e-mail: oscar.carlsson@midroc.com

A. W. Colombo et al. (eds.), *Industrial Cloud-Based Cyber-Physical Systems*,
DOI: 10.1007/978-3-319-05624-1_7, © Springer International Publishing Switzerland 2014

Fig. 7.1 LKAB plant for Fe-mineral processing

iron mine run by LKAB in north Sweden, with a focus on migration aspects. We demonstrate that the loose coupling provided by the SOA approach combined with the eventing capabilities of Event Driven Architecture (EDA) can benefit to both engineering, installation and maintenance of an industrial process control system, with the exception of hard real-time based control loops.

7.1 Introduction

The IMC-AESOP project has been investigating how a SOA would benefit to large-scale distributed systems in batch and process control applications. The project addresses in particular architectures where large number of service-compliant devices and systems distributed across a whole plant-wide system expose SCADA/DCS monitoring and control functions as services.

One essential investigated aspect is the cooperation between currently used synchronous DCS and SCADA and the new asynchronous SOA-based monitoring and control system, going beyond what the currently implemented control and monitoring systems typically deliver today. In this chapter, we will detail the development of an IMC-AESOP demonstrator at the premises of LKAB in Sweden (see Fig. 7.1), implementing an overall control scenario for an existing plant lubrication system and addressing the migration process from classical control systems to the new concepts addressed by the project.

Lubrication systems are typical critical systems for almost all process industries. The lubrication control system provides important information that can be used by operators, maintenance department, planning department and management to avoid critical and damaging incidents, improve production and plant efficiency, analyse the state of the system and implement predictive maintenance.

The IMC-AESOP plant lubrication use case addresses a number of key aspects of the project, such as enabling SOA on low level devices, SOA in closed-loop control, integration into an actual plant environment and migration from a scan-based PLC to an event-based SOA system. The advantages of using SOA-based solutions for industrial process monitoring and control, as well as their impact have been already recognised [3, 11, 13] and an architecture has been proposed [14], including the key relevant technologies [9].

In the following sections, we first outline the existing control and monitoring system, after which we describe the proposed architecture and components in-line with the overall IMC-AESOP vision [9, 14]. Subsequently, we provide the main implementation details and aspects of migration into a SOA-based solution. Finally, we summarise the results of the validation in the real plant.

7.2 Prototype Architecture

The lubrication system, shown in Fig. 7.2, is deployed in the LKAB pelletizing plant[1] on a number of independent systems which have limited data exchanges with the larger Distributed Control System (DCS), and from this perspective they behave as black boxes. One of these systems will be migrated from the current implementation using a PLC to a SOA system.

Similar migration efforts are described in [6] where XML/DPWS is used exclusively as a SOA implementation. In order to extend the service approach to highly constrained embedded devices we propose to use binary encoding for XML and the application protocols which is not investigate in the aforementioned work.

As shown in Fig. 7.2, the existing lubrication system includes two lubrication circuits controlled by a Programmable Logic Controller (PLC) receiving start/stop commands from a DCS. Each lubrication circuit is connected to a pump controlled by the DCS through a digital output. More than 70 AS-i (http://www.as-interface. net) position switches combined with various digital inputs are scanned periodically by the PLC to get fluid distribution status over each lubrication circuit. Based on this sensors information the PLC controls each pump and directs the fluid to the appropriate circuit. As mentioned above, there is a very limited communication with operational layer, although a touch panel provides a local supervision capability.

The prototype proposed for IMC-AESOP consists in replacing the existing PLC with an SOA-based system. Thus, the current PLC cabinet is replaced with a SOA-based cabinet and connected to a maintenance station (SCADA), as shown in Fig. 7.3.

---

[1] http://www.lkab.com/en/About-us/Overview/Operations-Areas/Kiruna/

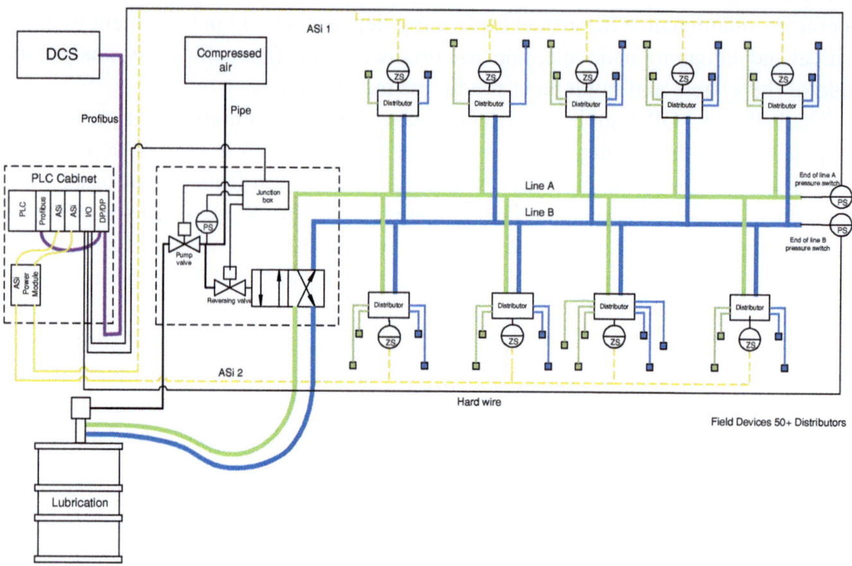

Fig. 7.2 Existing system

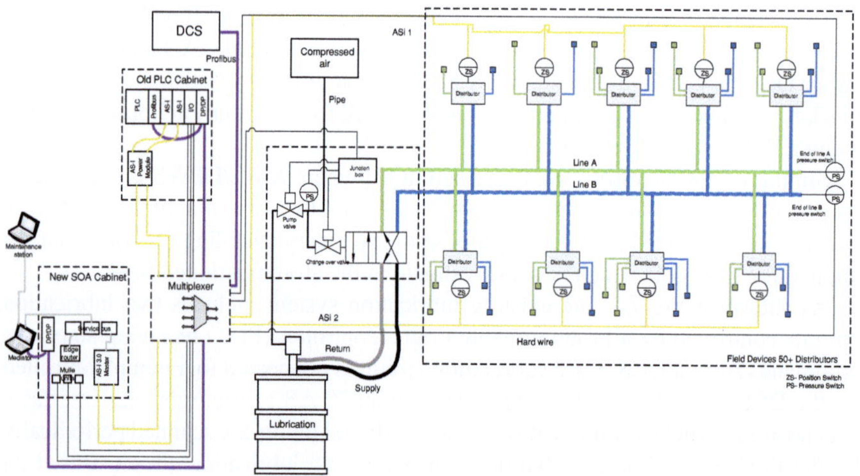

Fig. 7.3 Proposed prototype

7.3 SOA Components

The proposed SOA is shown in Fig. 7.4. Only the DCS part (dotted line) is inherited from the legacy system, other components, including the SCADA, are part of the SOA demonstrator. From top to bottom:

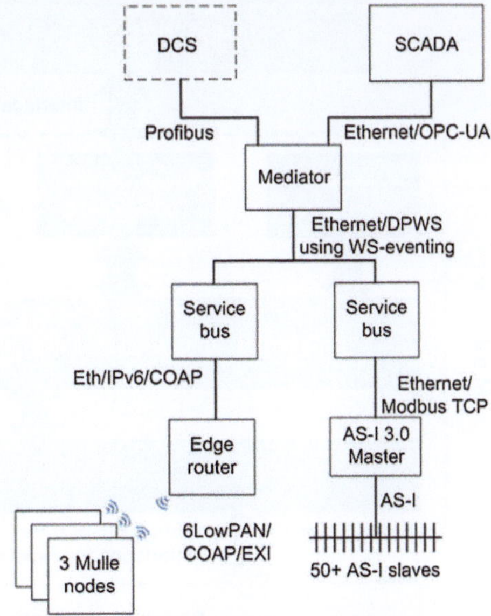

Fig. 7.4 Proposed prototype service-oriented architecture

- The SCADA provides advanced local control and monitoring capabilities;
- The Mediator provides an interface between the device layer (with the DPWS protocol) and the control and supervision layer (both Profibus and OPC UA protocols);
- The Service Bus provides an abstract layer on top of field devices, providing both synchronous and asynchronous data collection mechanisms. It implements as well the control logic that was originally running from the PLC;
- Wireless sensors and actuators (Mulle nodes and edge router) provide process I/O facilities.

7.3.1 SCADA

To replace and extend the HMI functionality provided in the legacy system by an integrated touch panel connected to the PLC, a commercially available SCADA solution was used and configured for the use case. The solution used provides a flexible way of presenting data and configuring the system parameters.

Using an OPC UA client, accessing the server provided by the Mediator, the system can be accessed from anywhere on the connected network as opposed to the current local access restriction. At the same time, the OPC UA server provides a flexible way to access the system with other standardised tools providing a wide array of possibilities.

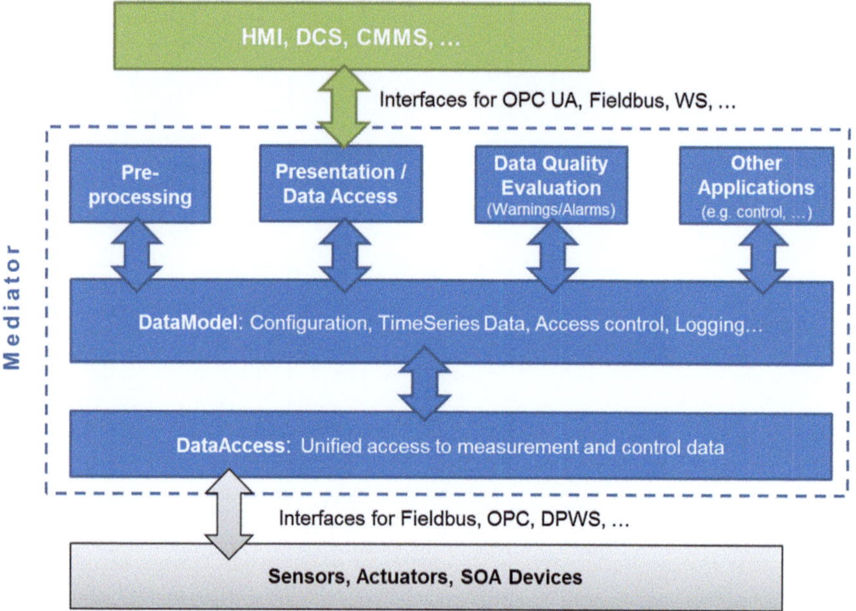

Fig. 7.5 Mediator structure

7.3.2 Mediator

The Mediator provides a runtime system for monitoring and controlling of process facilities by integrating both legacy as well as SOA-based technologies [12]. It has been built based on an actor-based middleware for fault-tolerant, distributed SCADA systems [8]. The adoption of the actor model [7] for the Mediator implementation results in less complexity and increased reliability compared to conventional (thread-based) approaches to the programming of concurrent processes. As all relevant sub-systems are actors that interact with each other only by message passing without sharing common data structures, the actor-based design of the Mediator also greatly simplifies the distribution of parts of the Mediator system.

Figure 7.5 shows the basic structure of the Mediator. Its core part consists of a data model that describes the logical view of the monitored facilities and also contains all relevant information for acquiring data including communication. The Mediator communicates with the Service Bus through DPWS and also supports basic authentication over SOAP.

For the integration of different communication protocols and information models of various devices and other data sources, an abstract data access layer has been introduced. By providing adapters implemented as actors, any required protocol can be integrated. For the application described in this chapter, the PROFIBUS protocol (for connecting to the DCS) as well as the DPWS protocol (for connecting to the

Service Bus) has been implemented. In a similar fashion, any processing of the data for pre-processing, control, KPI calculation or presentation to the SCADA HMI layer is easily extendable by providing appropriate adapters.

Within the framework of this SOA system described above, the Mediator data model (including alarms) is presented to the HMI of the maintenance application using the OPC UA protocol. The Mediator software is implemented using the Scala programming language (http://www.scala-lang.org) and therefore requires a Java Virtual Machine (JVM) at runtime. For the realisation of the actor system, the Scala-based library Akka (http://www.akka.io) is used, which is designed for building highly concurrent, distributed, and fault tolerant event-driven applications on the JVM. It is developed using Scala but can also be used in Java.

Akka actors efficiently implement the actor model. Therefore, the resource usage of individual actor instances is very small, allowing the creation of more than 2 million actors per GB of RAM [19]. The DPWS protocol has been implemented by using a customised version of the open source framework WS4D-JMEDS. For the OPC UA server adapter implementation the .NET-based OPC UA stack of the OPC Foundation was used. Because of this, the Mediator software must run on a Windows-based PC. In this use case, an industrial PC was used with 2.20 GHz Intel Atom CPU and 2 GB of RAM running Windows 7. The PROFIBUS connection to the DCS was realised using a USB connected PROFIBUS master that was controlled by the Mediator software via OPC.

7.3.3 Service Bus

In complement to the Mediator, the Distributed Service Bus provides an additional integration of heterogeneous systems supporting various communication media, protocols and data models. Such integration is enabled through loose coupling-based protocol connectors. Each protocol connector connects devices and services into the DSB data model representation. Thus, the Service Bus provides, through a defined abstract layer, a common representation of those devices and services.

In this demonstrator:

- A Modbus connector is used to connect to the AS-I subsystem through the Modbus to AS-I gateway;
- A CoAP connector is used to connect to the CoAP subsystem through the CoAP edge router;
- A DPWS connector is used to connect to the Mediator.

The distribution feature provided by the DSB is particularly suited to the management of large-scale distributed systems, and to the distributed nature of this demonstrator in particular. As shown in Fig. 7.6, one node of the Distributed Service Bus handles both the wireless nodes while the other node handles the AS-I sensors. Different communication technologies and different quality of service requirements

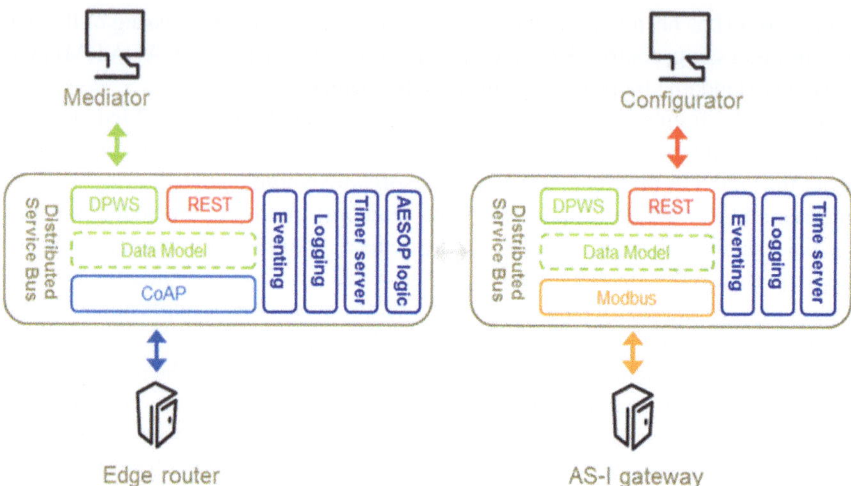

Fig. 7.6 Distributed service bus architecture

are served by different nodes. The wireless subsystem for instance handles several control functions (start, stop…) which are more critical than AS-I sensors information. The distributed architecture allows uncoupling those two subsystems.

The Service Bus has been implemented on two Raspberry Pi devices running Linux operating system and featuring 512 MB of RAM and 700 MHz ARM CPUs. As shown in Fig. 7.6 the main software components of the Service Bus are a pivot data format, a set of connectors acting as external interfaces (DPWS, REST, CoAP and Modbus), an eventing module, a time synchronisation (PTP) module, a logging (syslog) module and the IMC-AESOP logic which is reproducing the application logic from the existing PLC.

The two instances of Service Bus dynamically discover each other at startup with WS-Discovery and rely on DPWS for message exchanges between them. A basic cyber-security protection is provided by the combination of Role Base Access Control (RBAC) and user authentication mechanisms.

7.3.4 Wireless Sensors and Actuators

The recent use of internet protocols and web technologies for distributed sensor network installations is gaining wider acceptance [18]. The Wireless Sensor Actuator Network (WSAN), i.e. an industrial approach to Internet of Things (IoT) [2], is built on the 868MHz version of the IEEE 802.15.4 radio standard, which enables low-power communication through concrete walls and long-range communication at line of sight operation. The use of Industrial Internet of Things (IIoT) is suitable to use in combination with lightweight embedded systems that are used to measure (and control) physical parameters of interest.

To make the system scalable and integrate with the IMC-AESOP service cloud, IPv6 was chosen as network protocol. To make the IPv6 network layer comply with the IEEE 802.15.4 Link layer, the 6LoWPAN adaptation layer is used, 6LoWPAN compresses and reduces the data overhead so less energy is required to transfer the information between wireless nodes. IPv6 also enables unique identification of every sensor node using 128bit IPv6-address. The use of IPv6 also by default includes the network layer security feature of IPsec.

Figure 7.4 shows the edge router which performs translation between IPv4 over Ethernet and IPv6 over 6LoWPAN (IEEE 802.15.4) networks. The edge router also hosts time synchronisation services (NTP and PTP) and CoAP services such as data proxy, and also logs the performance of the WSAN. CoAP is a protocol designed for scalability and simplicity [1], whilst being backwards compatible with the much used HTTP protocol.

Mulle devices (http://www.eistec.se) serve as I/O nodes connecting lubrication pressure switches, air pressure switches, pump valves, reversing valves and indication lights. Mulle nodes communicate using Efficient XML Interchange (EXI) (http://www.w3.org/XML/EXI) and CoAP on top of 6LoWPAN. The services hosted by the Mulles support input, output, filtering, logging and configuration services. All the data are EXI encoded and transmitted using CoAP over 6LoWPAN.

Representing the information measured by the sensors in an efficient yet self-explanatory way is desirable. As the bandwidth in the wireless sensor network is limited, and the energy available in each sensor node is also limited, the efficiency parameter needs extra attention. The concept of SOA is highly interesting in this context as each measured parameter can be represented as a service to the other nodes, but also globally, as the sensors are connected to the Internet using IPv6.

7.3.4.1 I/O Nodes

In the demonstration setup, a total of 14 CoAP services (4 actuators, 6 sensors and 4 outputs used to indicate system status) were implemented. These were located on three different sensor nodes, each executing on a Mulle v6.2 [4, 5], equipped with a M16C/62P MCU running at 10 MHz and an 868 MHz low-power IEEE 802.15.4 transceiver.

The software on Mulles was implemented on Contiki, with built-in support for CoAP and 6LoWPAN; support for EXI was added to decrease the size of CoAP packets, which allowed us to avoid fragmentation of CoAP packets and improve robustness of communication. The clock of each node was synchronised to the clock of the edge router using the NTP protocol. In order to improve the time synchronisation performance, the solution proposed in [15] is an interesting approach that needs to be further investigated. The complete communication stack is shown in Fig. 7.7. To enhance the system's security, IPsec is planned to be deployed on the WSAN

Fig. 7.7 End node
communication stack

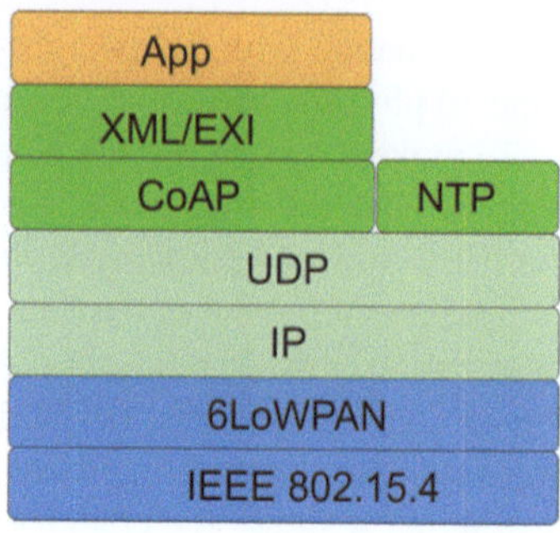

architecture as well. The use of IPsec on Contiki and 6LoWPAN has already been demonstrated [17].

7.3.4.2 CoAP and EXI

To implement a SOA-concept in a low-bandwidth 6LoWPAN WSAN used in this application, an efficient compression of the text-based XML service description and data is required. For this purpose EXI was used to represent the XML-based information in binary data format, this reduces the amount of bytes required to represent and transfer the service information. At the application layer, CoAP is used, as it is designed for resource constrained devices like the WSAN nodes used in the demonstration setup.

A key component of the migration of legacy systems to SOA is the use of standard and globally accepted formats for representing the exchanged information. One important result of this demonstration is that it is possible to use EXI for integration of sensor and actuator devices with the SOA automation infrastructure. This enables the implementation of RESTful Web services based on CoAP and EXI for industrial application with moderate real-time requirements.

7.4 Migration Aspects

Migration [4, 5] of a large DCS into SOA can be initiated with a smaller step where some key functionality is migrated and the basis of a SOA infrastructure is established in a part of the plant. A key aspect of the first step in a migration is to provide a platform for integration of more systems and functionality as the migration progresses. As such a first step this use case provides an interesting example of how a relatively simple system such as the PLC can be migrated and with the SOA infrastructure provide a possibility to connect different systems using a number of protocols without disturbing the functionality of the existing system.

This use case provides an example of migration of all required functional aspects which have been identified in the existing system and provides a minimum requirement of functionality for the SOA-enabled system. Most significant of these are:

- *Local control loop.* In the existing system, local control is performed within the PLC using internal timers and the pressure switches distributed throughout the system to trigger the start and stop of the lubrication pump and activation of solenoid valves. In the IMC-AESOP use case the functionality of the local control loop is assigned primarily to the Service Bus, accessing the CoAP services provided by the Mulle nodes for sensing and actuating. The main advantage of the SOA design is to provide added monitoring capabilities on the control loop (timers and sensors data are available as services). The performance of the local control loop was considered one of the more challenging aspects of the migration as it involved going from a high performance PLC, with synchronous polling of all devices, to a distributed system with asynchronous event-based communication.
- *Inter-protocol communication.* In the existing system, there are two communication protocols involved: The communication to the DCS is handled through Profibus and the collection of data from field devices is handled through AS-i. In the demonstrator, several new protocols are introduced as part of the architecture to allow communication within the SOA system, while the existing communication interfaces remain accessible through commercially available AS-i and Profibus master modules, respectively. The conversion between different protocols is handled by the Service Bus and the Mediator, as previously described.
- *Alarms and warning.* In the existing system alarms are handled through lists of Fault- and Reset-bits with a corresponding list of alarm texts, both in the PLC. In the SOA solution, those alarms are implemented as events collected from the alarm sources and brokered by the Service Bus. Any interested party can then subscribe to those alarms from the Service Bus. In the demonstrator, the SCADA, the DCS (both through the Mediator) and the Service Bus Web client are subscribers of process level alarms. Polling-based alarms remain available, which is particularly interesting in a migration context.
- *Operator manual override and Operator configuration.* Operator manual override and Operator configuration are the two key functionalities provided by the touch panel HMI in the existing system. In the SOA alternative, the Service Bus is exposing those two functionalities as services that can be called by any (authenticated and authorised) client application. In the demonstrator two client applications are consuming those services: the SCADA (through the Mediator) and the Service Bus Web client. As mentioned before, the loose coupling provided by this approach can be leveraged in future maintenance operations by allowing replacing transparently and independently either the server or the client part of those services.

Table 7.1 Time measurements

Event	Node	Time offset (ms)
End of line pressure switch	Mulle (sensor)	0
	Edge router	11
	Service bus	13
	Mediator	21

7.5 Validation Results

7.5.1 Functional Assessment

The functional validation of the overall architecture was performed on-site during a scheduled maintenance break of the plant. The IMC-AESOP prototype was connected to the lubrication system, by switching from the normally used operating cabinet to the new SOA cabinet. The lubrication system was then run for several hours to validate the functional behaviour of the prototype and collect performance data.

7.5.2 Performance Assessment

In order to measure the overall performance of the prototype, the components of the SOA synchronised their time using the PTP protocol (IEEE 1588). All the components were configured to send their logs to a centralised syslog server (IETF RFC 5424) for timing analysis. Table 7.1 summarises the average time it takes for a End of line pressure switch event to propagate from the Mulle device to the Mediator through the Edge Router and the Service Bus.

In this example, the CoAP Edge Router receives the event 11 ms after the Mulle detected the end of line pressure switch, then the Service Bus acknowledges the event 2 ms later and finally the Mediator 8 ms later. The total transmission time between the sensor (Mulle) and the Mediator is 21 ms which is above the current PLC cycle time but stays compatible with the application requirements.

7.5.3 Wireless Assessment

One parameter of interest that is important for successful deployment of 6LoWPAN devices is the size of the messages that the devices must exchange. Using XML is beneficial for integration of the devices with the data models and message formats used in the upper layers of the automation. By using EXI in strict XML schema mode for the low-bandwidth wireless links, the size of the XML messages is reduced more than 20 times. With that, the size of an EXI encoded digital IO process value with timestamp and quality indicator is 10 bytes as compared to 228 bytes for its plain XML counterpart. Another key performance indicator for wireless applications, especially in noise industrial environments, is the occurrence of retransmissions of packets. A retransmission wastes link bandwidth uses energy and increase latency. During the tests, retransmissions were at a low level, with a stable wireless network as a result.

7.5.4 Data Modelling

Enabling interoperability of the service specifications and data models is a key technological challenge that SOA systems are aimed to resolve. The full interoperability requires that the syntax and semantic service descriptions are well defined, unambiguous and enable dynamic discovery and composition. Thus far, most if not all SOA installations are enabling pure syntax interoperability with little or no support for standard-based semantic descriptions. The use of structured data formats only partially resolves the problem by supplementing the exchanged data with meta-information in the form of tags and attributes in the case of XML/EXI for example. The tag names are ambiguous and usually insufficient to describe the service functionality in full.

Applying application level data model standards is often used as a solution to that problem as the syntax to semantics mapping is predefined. Example of such standard is Smart Energy Profile 2 that clearly states the physical meaning of the tag names and structures defined for the service messages in the domain of energy management. One problem when complying with such standards is that they are almost always domain specific which requires mapping of the semantic descriptions from one standard to all others in use.

Another approach is to define generic semantic data model that is applicable to wide range of use cases. This is the approach selected for the work presented in this paper. The initial investigation highlighted the Sensor Model Language (SensorML) [16] developed by Open Geospatial Consortium (OGC) as a promising specification for generic semantic description of sensory data. However, the complexity and size of SensorML specification limit its use to more capable devices. Small-scale experiments with a number of sample SensorML messages showed that even EXI representation will not be sufficiently small to fit a battery powered wireless sensor nodes that have low-power and low-bandwidth radios.

Another possible specification for sensor data is the Sensor Markup Language (SenML) [10]. It has a very simple design that is consistent with RESTful architecture and is targeted at resource-constrained devices. The evaluation of SenML specification showed that it meets the requirements for hardware utilisation but there are areas that are too much simplified and insufficient to describe the data in the details required by the target application. Example of such limitation is the precision of the time stamping of the sensor data—SenML allows for up to seconds resolution that is not enough for most use cases. This led to the use of custom generic data representation that is reusing many of the design choices in SenML.

7.5.5 *Overall Assessment*

A general drawback of the proposed solution is obviously its lack of maturity, in a sense that it consisted of a set of prototypes provided by different partners, none of them being productised yet. This translated into both unreliability issues and integration complexity. Part of the integration difficulties consisted in having a specific configuration and monitoring interface for each partner component.

This heterogeneity and relative complexity of the demonstrator can in turn be perceived as an opportunity to validate the SOA approach, each component of the architecture exposing and consuming services to/from other components, with a fairly high level of loose coupling.

In a productised version of the demonstrator, all middleware components (Mediator, Service Bus, Edge Router and potentially AS-I gateway) would ideally be merged into one product, thus reducing the main complexity of the system. However, a projection to a productised version of the SOA middleware would still lead to a higher level of internal complexity compared with a less versatile PLC-based solution.

The main benefit of the proposed solution, compared with the installed solution, is to facilitate the overall system installation and maintenance. Although the installation benefit was not obvious on the demonstrator due to the multiplicity of technologies (and partners) involved, the maintenance and monitoring value was fairly obvious thanks to the advanced monitoring capabilities provided by the added services and displayed through the SCADA (timers, sensors values, alarms…).

7.6 Conclusion

The on-site validation of the IMC-AESOP prototype provided very positive feedbacks considering that both functional and performance results were in line with customer expectations, combined with added supervision and control capabilities at the SCADA level. SOA proved to be valuable both at device and application level by providing a high level of loose coupling between the various components of the system. Eventing complemented nicely the service-based architecture by reducing the overall latency of the information flow. On the wireless side, the tests show that CoAP-based services over 6LoWPAN can be used for process monitoring and control applications with no low-latency requirements. More research is needed though in order to improve both scalability and robustness and minimise latency.

Acknowledgments The authors would like to thank the European Commission for their support, and the partners of the EU FP7 project IMC-AESOP (http://www.imc-aesop.eu) for the fruitful discussions.

References

1. Bormann C, Castellani AP, Shelby Z (2012) Coap: an application protocol for billions of tiny internet nodes. IEEE Internet Comput 16(2):62–67. http://doi.ieeecomputersociety.org/10.1109/MIC.2012.29
2. Castellani A, Bui N, Casari P, Rossi M, Shelby Z, Zorzi M (2010) Architecture and protocols for the internet of things: a case study. In: 8th IEEE international conference on pervasive computing and communications workshops (PERCOM workshops), 2010, pp 678–683. doi:10.1109/PERCOMW.2010.5470520
3. Colombo A, Karnouskos S, Bangemann T (2013) A system of systems view on collaborative industrial automation. In: IEEE international conference on industrial technology (ICIT 2013), pp 1968–1975. doi:10.1109/ICIT.2013.6505980
4. Delsing J, Eliasson J, Kyusakov R, Colombo AW, Jammes F, Nessaether J, Karnouskos S, Diedrich C (2011) A migration approach towards a SOA-based next generation process control and monitoring. In: 37th annual conference of the IEEE industrial electronics society (IECON 2011), Melbourne, Australia
5. Delsing J, Rosenqvist F, Carlsson O, Colombo AW, Bangemann T (2012) Migration of industrial process control systems into service oriented architecture. In: 38th annual conference of the IEEE industrial electronics society (IECON 2012), Montréal, Canada
6. Feldhorst S, Libert S, ten Hompel M, Krumm H (2009) Integration of a legacy automation system into a SOA for devices. In: IEEE conference on emerging technologies factory automation (ETFA 2009). pp 1–8. doi:10.1109/ETFA.2009.5347068
7. Hewitt C, Bishop P, Steiger R (1973) A universal modular actor formalism for artificial intelligence. In: Proceedings of the 3rd international joint conference on artificial intelligence, Morgan (IJCAI'73), pp 235–245. Kaufmann Publishers Inc., San Francisco, CA, USA. http://dl.acm.org/citation.cfm?id=1624775.1624804
8. Hübner C, Thron M, Alex J, Bangemann T, (2013) Aktor-basierte middleware-plattform für fehlertolerante, verteilte scada-systeme. In: AUTOMATION, 2013, VDI Wissensforum GmbH, Baden-Baden, VDI
9. Jammes F, Bony B, Nappey P, Colombo AW, Delsing J, Eliasson J, Kyusakov R, Karnouskos S, Stluka P, Tilly M (2012) Technologies for SOA-based distributed large scale process monitoring and control systems. In: 38th annual conference of the IEEE industrial electronics society (IECON 2012), Montréal, Canada
10. Jennings C, Shelby Z, Arkko J (2013) Media types for sensor markup language (SENML). Technical report, IETF Secretariat. http://tools.ietf.org/html/draft-jennings-senml-10
11. Karnouskos S, Colombo AW (2011) Architecting the next generation of service-based SCADA/DCS system of systems. In: 37th annual conference of the IEEE industrial electronics society (IECON 2011), Melbourne, Australia
12. Karnouskos S, Bangemann T, Diedrich C (2009) Integration of legacy devices in the future SOA-based factory. In: 13th IFAC symposium on information control problems in manufacturing (INCOM), Moscow, Russia
13. Karnouskos S, Colombo AW, Jammes F, Delsing J, Bangemann T (2010) Towards an architecture for service-oriented process monitoring and control. In: 36th annual conference of the IEEE industrial electronics society (IECON 2010), Phoenix, AZ
14. Karnouskos S, Colombo AW, Bangemann T, Manninen K, Camp R, Tilly M, Stluka P, Jammes F, Delsing J, Eliasson J (2012) A SOA-based architecture for empowering future collaborative

cloud-based industrial automation. In: 38th annual conference of the IEEE industrial electronics society (IECON 2012), Montréal, Canada
15. Kim K, Lee SW, geun Park D, Lee BC (2009) Ptp interworking 802.15.4 using 6lowpan. In: 11th International conference on advanced communication technology (ICACT 2009), vol 01, pp 873–876
16. OCG (2007) Sensor model language (SensorML) implementation specification. http://www.opengeospatial.org/standards/sensorml
17. Raza S, Duquennoy S, Höglund J, Roedig U, Voigt T (2012) Secure communication for the internet of things—a comparison of link-layer security and ipsec for 6lowpan. Secur Commun Netw. doi:10.1002/sec.406, http://dx.doi.org/10.1002/sec.406
18. Shelby Z (2010) Embedded web services. Wirel Commun 17(6):52–57. doi:10.1109/MWC.2010.5675778, http://dx.doi.org/10.1109/MWC.2010.5675778
19. TypeSafe Inc (2013) Akka documentation, release 2.1.2. http://www.akka.io/docs/

Chapter 8
Implementing Circulating Oil Lubrication Systems Based on the IMC-AESOP Architecture

Roberto Camp and Andrei Lobov

Abstract Current circulating oil lubrication systems used in the process industry, specifically those used in the Pulp and Paper industry are still quite behind in terms of technological advances. These systems still rely on first or second generation SCADA systems, and utilise old and convoluted communication systems that rely on fieldbuses. High demand in quality and scalability is pushing for the use of SOA-oriented systems at all levels of large-scale process systems. This chapter focuses on how this approach affects this particular domain.

8.1 Introduction

Fluid automation plays an important role in large-scale industrial machinery. Hydraulic control is often used in industrial applications where electrical drives cannot provide enough power. Similarly, machines that highly depend on mechanical components require the use of lubrication that can more easily be achieved by fluid automation systems. One particularly important type of process found in fluid automation is the circulating oil lubrication process, which is required in pulp and paper, steel and oil and gas industries, to name just a few.

The operation of oil circulating lubrication systems in the context of large-scale distributed systems is particularly challenging. The amount of lubrication required in these massive machines, the communication infrastructure, and the requirement to comply with strict environmental regulations are just some of these challenges. Some of these challenges can be addressed by implementing new technologies in

R. Camp (✉)
FluidHouse, Jyväskylä, Finland
e-mail: roberto.camp@fluidhouse.fi

A. Lobov
Tampere University of Technology, Tampere, Finland
e-mail: andrei.lobov@tut.fi

A. W. Colombo et al. (eds.), *Industrial Cloud-Based Cyber-Physical Systems*,
DOI: 10.1007/978-3-319-05624-1_8, © Springer International Publishing Switzerland 2014

the communication infrastructure of these systems that could enable the adoption of advanced monitoring techniques (and systems), of both oil quality and massive amounts of lubrication points. This will reduce the costs (both environmental and production) associated with the cost of oil and the maintenance of the machines they lubricate. The following sections will briefly elaborate on the important aspects that need to be taken into consideration in circulating oil lubrication systems.

8.1.1 The Importance of Lubrication

The objective of oil lubrication systems is to ensure the constant lubrication of moving mechanical parts. Even though this basic objective is quite simple and straightforward, these systems are critical to manufacturing processes due to the fact that they prevent the deterioration of most of the mechanical components of machines and ensure constant and reliable operation. Additionally, the main contributor for mechanical part degradation in hydraulic lubrication systems is the contamination of the lubricant used.

The existence of friction, dirt particles in the lubricant, or water in the case of synthetic oils cause wear and accelerate the degradation that eventually lead to unavailability, shutdown and ultimately to machine breakdown [3, 4]. In the specific case of paper mills the mechanical components of most importance can be said to be the bearings. In Ref. [2], a study of the different types of paper machine bearing failures concludes that 34.4 % of the failures are due to inadequate lubrication.

Additionally, a study realised by Ref. [7] of the Massachusetts Institute of Technology and presented by Ref. [3] studied the major reasons for component replacement. It was concluded that 70 % of the necessary replacements was caused by surface degradation, 50 % of which was caused by mechanical wear, and 20 % caused by corrosion due to the water contamination in mineral oils. A more complete representation of this study can be seen in Fig. 8.1. This highlights not only the importance of lubrication systems themselves, but also the importance behind the monitoring of oil quality [1].

8.1.2 Lubrication in Paper Machines

Lubrication systems in paper machines are somewhat particular, not in the method of lubrication, but in the amount of lubrication they require. Any typical paper machine consists of dozens of bearings and mechanical components that are in constant movement and require lubrication in order to avoid overheating and damage. Additionally, paper machines are considerably large, capable of being as large as 100 m long, 20 m tall (occupying two floors) and around 20 m wide. This creates the need to lubricate anything between 600 to 1,200 points in a paper machine. Figure 8.2 shows the side view of a paper machine and represents all the different cylinders and bearings it has.

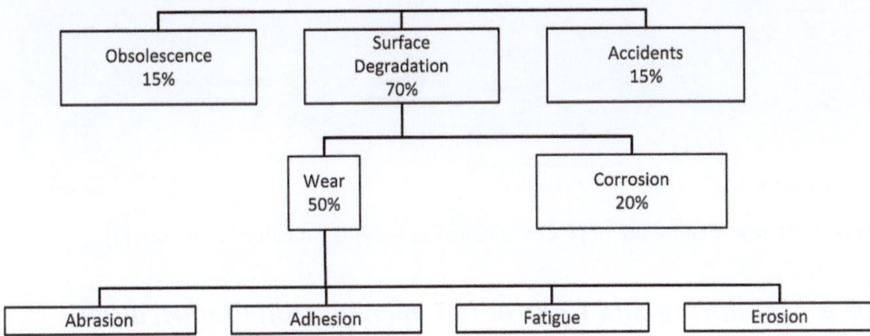

Fig. 8.1 Reasons for replacing components [3]

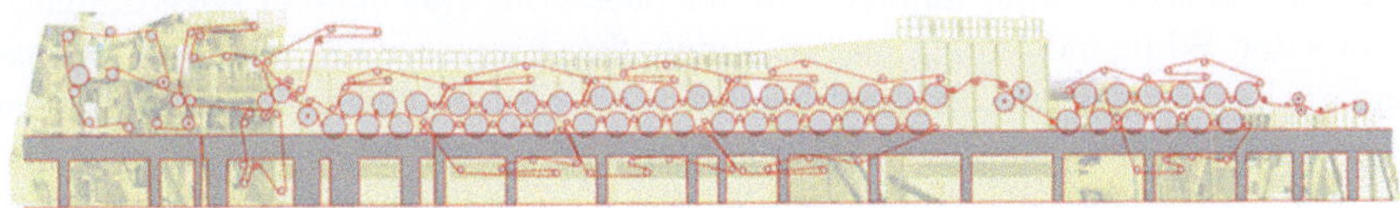

Fig. 8.2 Paper machine side view

In order to lubricate all these points, paper machines typically use so-called circulating oil lubrication systems. These systems utilise one or more lubrication oil reservoirs that pump oil to measuring stations. These stations can contain anything between 1 and 100 flow metres (although usually ranging between 20 and 50), each of which measures the flow rate at which a lubrication point receives lubrication oil. After each point is lubricated, the oil is caught in platters located below each lubrication point and is returned to the reservoir by gravity. The oil is then cleaned, cooled and filtered before being pumped back into the paper machine for more lubrication. Some more modern systems also remove air bubbles from the oil by using centrifugal force.

Depending on the size of the paper machine, measuring stations are placed on the paper machine in different locations. Each of these measuring stations will have a certain amount of flow metres which depends on the amount of lubrication points in that particular section of the paper machine. The type of flow metres used may vary, but it is typical to use positive displacement flow metres. These flow metres measure the amount of volume passing through them by counting the pulses generated by the flow of oil. These flow metres come in different sizes and depend on the lubrication point they are connected to, their physical size is also directly related to the amount of flow range they can measure. Typical flow metres in paper machines allow flow rates ranging between 0.2 and 20 L/min. Occasionally, flow metres that have a bigger flow range (up to 50 L/min) can be found in locations where special gearboxes are found.

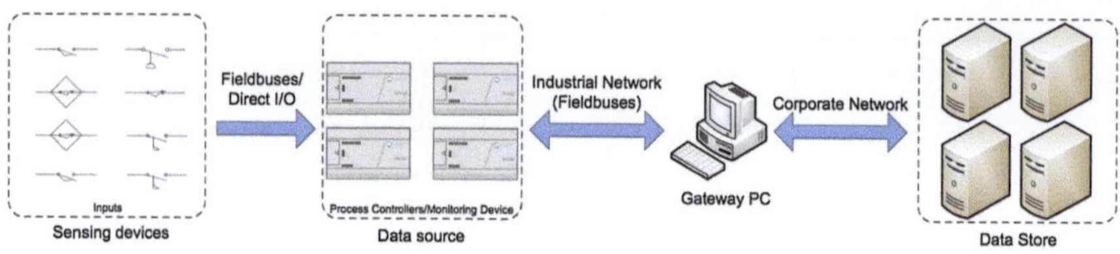

Fig. 8.3 Typical architecture of process monitoring systems—modified from Ref. [5]

8.1.3 Monitoring and Control in Paper Machine Lubrication

While many different types of monitoring systems have been implemented throughout the years; most of these are tailored to the manufacturing system/process it is designed to monitor. While monitoring itself is simply the observance and analysis of a system's behaviour, current condition monitoring systems require that such information is readily accessible from both near and far. The former refers to local monitoring stations, and the latter to remote control rooms and corporate level monitoring interfaces. This is commonly accomplished by implementing fieldbus communication networks and systems capable of using such networks. In the end, this leads to a common pattern followed in the implementation of these monitoring systems, a generic and common architecture can be seen in Fig. 8.3.

As it can be seen in Fig. 8.3, a typical industrial monitoring system has three components: A data source, a communication infrastructure and a data store [5]. This is the normal approach to monitoring systems because it is the most intuitive and straightforward approach, mainly due to the fact that all systems are different and there is no common standard to implement monitoring systems. In most systems all the information can be obtained from the controlling entity, namely the process controllers.

The advantages to this can be many. For instance, if the desired variables are already being measured, there no need to install additional sensors; however, there is a need to programme routines that handle the information gathering. Additionally, there is a need to adapt and use the communication infrastructure to be able to obtain the information gathered by the data source. Once the information has been gathered in a conventional information system it can be stored in any given format according to the system existing standards.

Circulating oil flow monitoring systems are not very different from the generic systems previously mentioned. The varying factor is mainly the way a monitoring system is integrated or designed in conjunction with the oil flow monitoring system. Hence, the way the monitoring system is designed and implemented depends on the manufacturer and on the requirements of the system or factory where they are installed. It is therefore no surprise that they are all different and that academic information regarding these systems is practically non-existent. There is, however plenty of commercial solutions that are specifically designed to deal with circulating oil monitoring systems. While some of these are better than others, they still follow the same pattern shown in Fig. 8.3.

8.1.4 Requirements and Motivation

The previous subsections have briefly introduced the domain and the subject matter of the oil lubrication use case for the IMC-AESOP project. Given the state of the art of these systems, and the original objectives and the requirements defined in the early stages of this project, the use case demonstrator presented in this chapter had as its original objectives to:

- To introduce the SOA paradigm into circulating oil lubrication systems.
- To prove the feasibility of implementing CEP systems in oil lubrication monitoring systems.
- To evaluate the behaviour of event-based monitoring in a domain that has traditionally been polling based.
- To validate the feasibility of using cloud-based services in oil lubrication systems.

8.2 Oil Lubrication System Description

The oil lubrication use case for the IMC-AESOP project addressed the manner in which lubrication systems in paper machine are monitored. As mentioned in the previous chapter, given that paper machines require the active lubrication of hundreds of different points, each of these points requires the existence of its own flow metre, and therefore its monitoring. In legacy systems flow metres are monitored either manually by the process operators, or by elaborate monitoring systems that depend on traditional fieldbuses, or only allow for a single entity (monitoring room/computer) to show the captured information, as mentioned in Sect. 8.1.3.

This type of system requires that the operators go separately to the flow metre and navigate the corresponding interface (if it exists) to be able to visualize the relevant flow rate. Alternatively, it requires the operator to go to the monitoring computer. This monitoring method, while in use, implies that the monitored information has to be either centralised or is completely unavailable. Additionally, the generated alarms are localised and grouped by station, which generally makes possible problems slow to identify. This use case implementation seeks to implement a SOA approach to the monitoring of lubrication systems in paper machines.

The oil lubrication demonstrator focuses on demonstrating the implementation of a SOA in circulating oil lubrication systems. Main components of such lubrication systems include the following, and can be seen from Fig. 8.4:

- The *Lubrication unit* is an oil reservoir that can hold thousands of litres of oil. This oil tank has integrated pumps that move the oil from the reservoir to the various measuring stations, which are attached to any type of paper machine. After the oil is distributed to bearings and lubrication points all over the paper machine, it returns (due to gravity) back to the oil reservoir. Here it is cleaned, filtered, cooled and then put back into the circulation system.

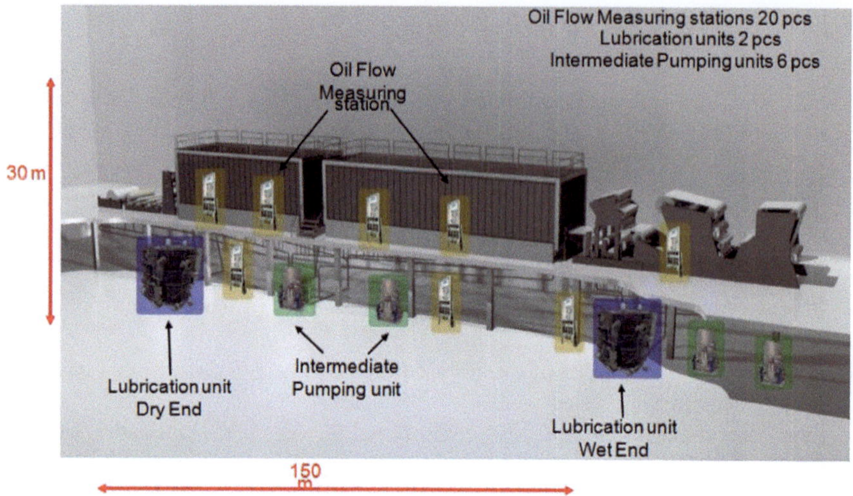

Fig. 8.4 Illustration of the main system components of a circulating oil lubrication system installed in a paper machine

- *Measuring stations* are panels with numerous flow metres mounted on them. The number of flow metres on each panel can range between 20 and 50, depending on where the flow panel is located in relation to the paper machine.
- *Intermediate pumping units* are used in case the distance between the lubrication units and the measuring stations is too long. They pump the oil to the distant measuring stations.
- The Ignition *SCADA system* has capability to subscribe to the Web service events, which are generated by the demonstrator.

The IMC-AESOP demonstrator consists of two parts: real and simulated equipment. The first part is a real measuring station with circulating oil, as in a real lubrication system. The second part is a software simulator emulating the whole lubrication system of a paper machine. The following sections describe each of the relevant parts of the demonstrator in the use case.

8.2.1 Real System

The real system (see Fig. 8.5) consists of a real measuring station equipped with positive displacement flow metres, which are connected to a DPWS embedded controller. This controller counts the pulses and produces flow rates by calculating the generated pulses per minute. After the calculation of the flow, its specific values are encapsulated in WS events that are generated every time a significant change of the flow rate is detected. The controller used in the real part of the demonstrator can be characterised as follows:

Fig. 8.5 Real system (demo implemented)

- The controller has a WSDL file stored in it, making it a discoverable device.
- The controller monitors 24 flow metres and their flow rate events.
- The controller has configuration Web services.
- The controller generates events for other values, such as temperature, pressure, etc.

8.2.2 Simulated System

This system provides the simulated behaviour of a measuring station. Up to 23 measuring stations can be simulated (see Fig. 8.6), each with the same characteristics as the real system, albeit simulated. There are two lubrication units simulated, one constantly running and another optional lubrication unit that can generate relevant events. Each activated measuring station and lubrication unit behaves as an individual discoverable device to which any client can subscribe and any of the configured Web services can be invoked. Main characteristics of the simulated system are as follows:

- Each simulated controller has a WSDL file stored in it, making it a discoverable device. This means up to 23 discoverable devices per running simulation.

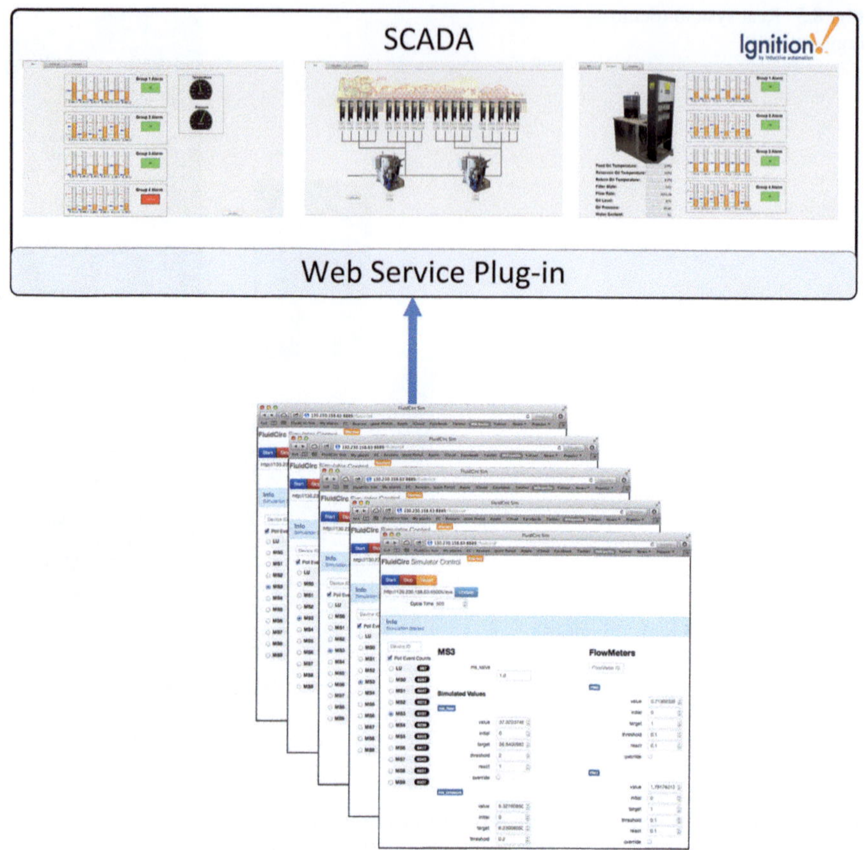

Fig. 8.6 Scheme of the simulated system

- Each simulated measuring station controller can have 24 or 48 flow change events.
- Each simulated measuring station controller has two configuration Web services.
- Each simulated measuring station has events for other values, such as temperature, pressure, etc.
- There are two simulated controllers that will represent the lubrication units.
- Many instances of the simulation can be run on the network.

8.2.3 Cloud Integration

Cloud integration is envisioned to make the CEP service available and remotely accessible (see Fig. 8.7). The role of the service deployment to the cloud is in the long-term reporting scenario. It is envisioned that the cloud infrastructure should primarily allow provision of additional computational resources and minimise the maintenance cost in the long run.

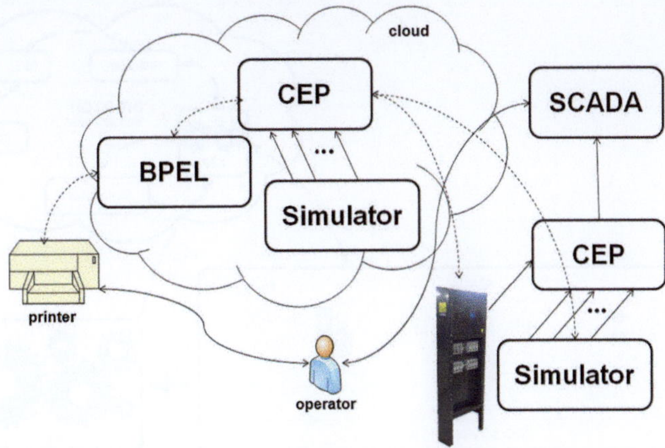

Fig. 8.7 Cloud integration

8.2.4 Scope of the Demonstration

The scope planned to consider a scenario that contains the following:

- 1 real measuring station.
- Up to 23 simulated measuring stations; each simulated station can have up to 48 flow metres.
- Up to 2 simulated lubrication units.

This gives the maximum number of $(20 \times 48) + (1 \times 24) = 984$ flow metres and events within the demonstration. Each flow metre has its own event indicating a flow change. Each simulated measuring station simulates also other variables: oil temperature, pressure, viscosity and possibly others.

8.2.5 Proposed Architecture

The system architecture proposed for the oil lubrication demo is shown in Fig. 8.8. It includes both the physical measurement station with associated controller (real system) and the simulator (simulated system). Both supply data directly to Ignition SCADA system and then also through DPWS to the CEP / StreamInsight engine.

8.3 Oil Lubrication System Implementation Details

In Sect. 8.2, the functional description of the oil lubrication use case was presented. This section will complement the previous description by describing the main components present in the use case from the architectural perspective. In some cases the

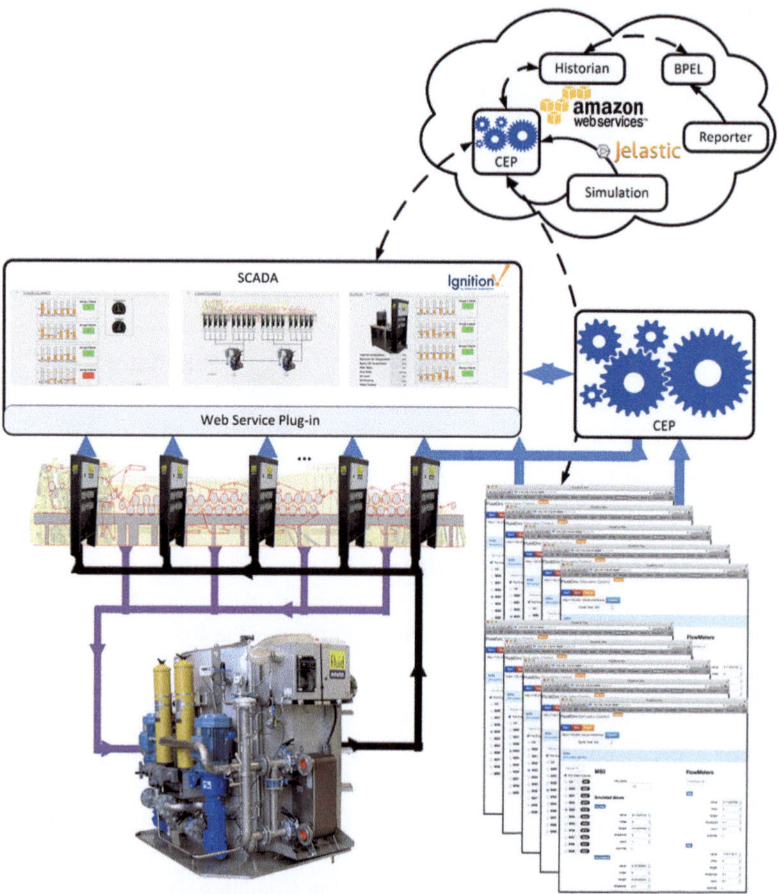

Fig. 8.8 Architecture of the oil lubrication demonstrator

architectural components presented will represent similar components to the ones presented in the functional description and the IMC-AESOP architecture. This is mostly because some components presented here cannot be encompassed by the other main components and need to be presented individually. The overall structure of the demonstrator, however, is based on the IMC-AESOP architecture.

8.3.1 Architectural Description

Figure 8.8 shows the main components in this architectural description; In general, at the lowest levels we can find the DPWS embedded devices, in parallel we can also see the simulation model, which represents both DPWS devices and the behaviour of oil lubrication systems. From this point, the devices and simulation can send information both to the CEP engine and directly to Ignition SCADA system, which interprets DPWS messages through its interface.

Moving further up, the 'Cloud' comes into play; running services and simulations on local or third party servers (in the cloud) the architecture provides business process orchestration (BPEL), other CEP engines, historian and reporting services (which are directly related to the IMC-AESOP architecture). More specifically the components are:

- *DPWS devices*. These are RTU devices that are programmable and can function as data pre-processing computers that can encapsulate information into SOAP-XML messages.
- *DPWS simulation system*. Same function as the DPWS devices, with the exception that a behavioural model which simulates the way real lubrication systems work is running in the background of this simulation.
- *CEP*. This system captures messages and post-processes data and information to create more complicated information. This information can later be provided to any system that desires to consume it.
- *Ignition*. The SCADA system that captures information sent by the CEP and the devices. It also enables the visualisation of information and monitoring.
- *CEP and Simulation (on the cloud)*. Has the same function as the previously described CEP and simulation system, they run on the cloud.
- *Reporter*. Runs on the cloud and enables the generation of reports, which can later be consumed by printers or other clients.
- *Historian*. Enables the management of historical data and information.
- *BPEL Orchestrator*. Business processes can be created while the service runs on the cloud.

8.3.2 Tools

It is important to consider that many of the tools used in the development and implementation of the oil lubrication use case, function as a framework. Some of these tools have small functions, some have more elaborate functions, but overall, all of them are needed in order to successfully implement the use case.

In general, for the implementation of this use case there are two different types of tools used. Development and deployment tools are used to create, programme and configure the necessary behaviours and functions of the use case. Additionally, testing tools are used to test the developed and deployed system.

The main development and deployment tools used are:

- *Apache ServiceMix*. A platform providing useful functionality for integrating technologies internal to components, and WS frameworks for exposing services.
- *Apache Camel*. An open-source integration framework based on enterprise integration patterns, which provides connectivity to a wide array of technologies/-transports/protocols/APIs, included in ServiceMix.

- *Jetty*. A simple HTTP server, which can be used for consuming and producing HTTP requests.
- *Web services for Devices—Java Multi-Edition DPWS Stack (WS4D–JMEDS)*. An open-source stack for developing DPWS clients, devices, and services.
- *Java API for XML Web Services (JAX-WS)*. A Java API for developing Web services.
- *Windows Communication Foundation (WCF)*. Windows run-time API for developing SOA applications in C#.
- *StreamInsight*. A platform for developing and deploying CEP applications from Microsoft
- *Ignition Server*. A commercial HMI/SCADA system with integrated OPC-UA server.
- *Ignition Developer API*. An application programming interface for developing custom modules for the Ignition Gateway, Designer, or Client in Eclipse.
- *Eclipse BPEL Designer*. A graphical editor for creating BPEL processes.
- *Orchestration Engine*. A tool developed at TUT for executing WS-BPEL processes.

The main testing tools used are:

- *WCFStormLite*. For testing WCF Services.
- *DPWS Explorer*. For testing services on DPWS devices.
- *UA Expert*. A free OPC-UA Client for testing OPC-UA and DPWS integration.

The tools presented were used in the development of the use case, and can be mapped to the architectural components of the IMC-AESOP architecture [6] as shown in Table 8.1.

8.4 Results

The results of the oil lubrication use case are to a certain degree self-evident in the monitoring features it provides as a result. However, one the most important questions that this use case was meant to answer was the possibility, and feasibility of SOA based monitoring in large-scale systems. This section of the chapter addresses the final results of the use case by briefly explaining the evaluation methods and the obtained results.

8.4.1 Tools and Measurement Setup

As mentioned previously, the use case aimed to implement a Service-Oriented Architecture in oil lubrication systems commonly used in paper machines where measurements are taken for hundreds of lubrication points.

Table 8.1 Use case tool to component mapping

Component	Platform	In cloud	Language	WS-Framework	Other tools
Reporting	ServiceMix	No	Java	JAX-WS	camel-printer, camel-mail
DPWS subscription proxy	ServiceMix	No	Java	WS4D	Jetty as HTTP server for browser config
SCADA	Ignition server	No	Java	WS4D	Ignition server
Simulator	ServiceMix	Both	Java	WS4D	Jetty as HTTP server for browser config
CEP	StreamInsight	Both	C#	WCF	
Orchestrator	Standalone	Yes	Java	Custom DPWS	BPEL designer plugin (Eclipse)
Historian	StreanInsight	Yes	Java	JAX-WS	SQLServer R2
Devices	S1000	No	ST	Custom DPWS	None

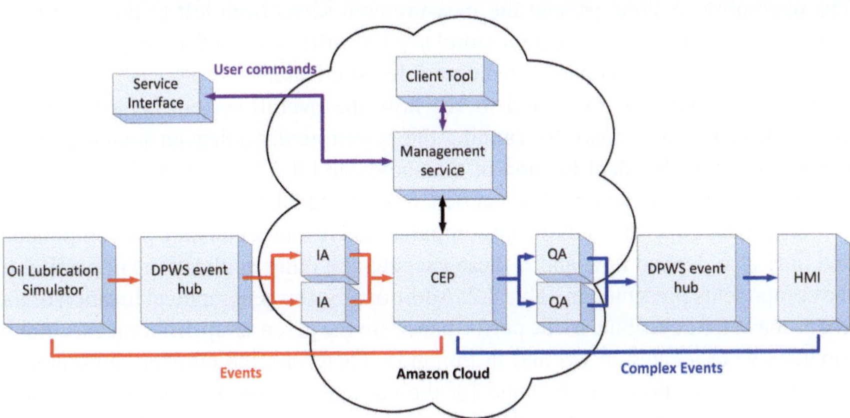

Fig. 8.9 Software components to assess oil lubrication for paper machines

The overall architecture for oil lubrication for paper machine use case shown in Fig. 8.9. On the left hand side, the oil lubrication simulator which is used to overwhelm the cloud-based CEP service with DPWS events by sending many events in a small period of time, this CEP outputs complex events that are generated by interpreting event data.

The CEP as a service gives a possibility to remotely (via Service Interface) add queries dynamically, this makes it possible to adapt to the needs of any applications where the potential queries may not be known at design time. In the case of the use case test environment, the Amazon cloud is used to host the CEP service.

Table 8.2 Component location and platform characteristics

Component	Location	Characteristics
Oil lubrication simulator	PC	OSX 10.8.5
		2 core
		2 GHz processor
EventHub		Memory 8 GB
		Finland
CEP service	Amazon elastic cloud computing (EC2)	m1.small[a]
		ECU5[b]
Management service		Low network performance (USA)
Complex event consumer	Jelastic PaaS	Application server: Glass fish 3.1.2.2, Six cloudlets[c]

[a] http://aws.amazon.com/ec2/instance-types/instance-details/
[b] One EC2 Compute Unit (ECU) provides the equivalent CPU capacity of a 1.0–1.2 GHz 2007 Opteron or 2007 Xeon processor
[c] A cloudlet is roughly equivalent to 128 MB RAM and 200 MHz CPU core

Each element of the chain shown in Fig. 8.9 can serve as a remote component. The upcoming sections present the measurement setup from left to the right from event generation at oil lubrication simulator until the arrival of a complex event to the complex event consumer hosted on the Jelastic PaaS.

The performance is evaluated to see how the overall system behaves once it experiences a 'heavy' load. By running the system under a heavier load that it can process, enables the identification of thresholds and the evaluation of applicability of the solution to the system that can benefit of using CEP.

Performance also depends on the computational resources hosting the components and may also depend how and if these execute any other applications in parallel to the components presented in Table 8.2. Additionally, the geographical location of the nodes may also contribute to the performance (In the given setup different executable modules were located in US and in Finland). Therefore, the numbers presented in the upcoming sections can be valid for the particular setup, but they may differ in some other cases. Nevertheless, these allowed us to find key points, which should be considered while using cloud-based applications for industrial applications of CEP.

8.4.1.1 Oil Lubrication Simulator

The oil lubrication simulator enables the simulation of the behaviour of oil lubrication systems in order to assess CEP performance. It is used to overwhelm the CEP service with events. Figure 8.10 shows a web interface of the oil lubrication simulator (thus, the simulator itself can be deployed remotely). The simulator represents oil lubrication systems for paper machines. For the purpose of the CEP test the oil lubrication system is composed of:

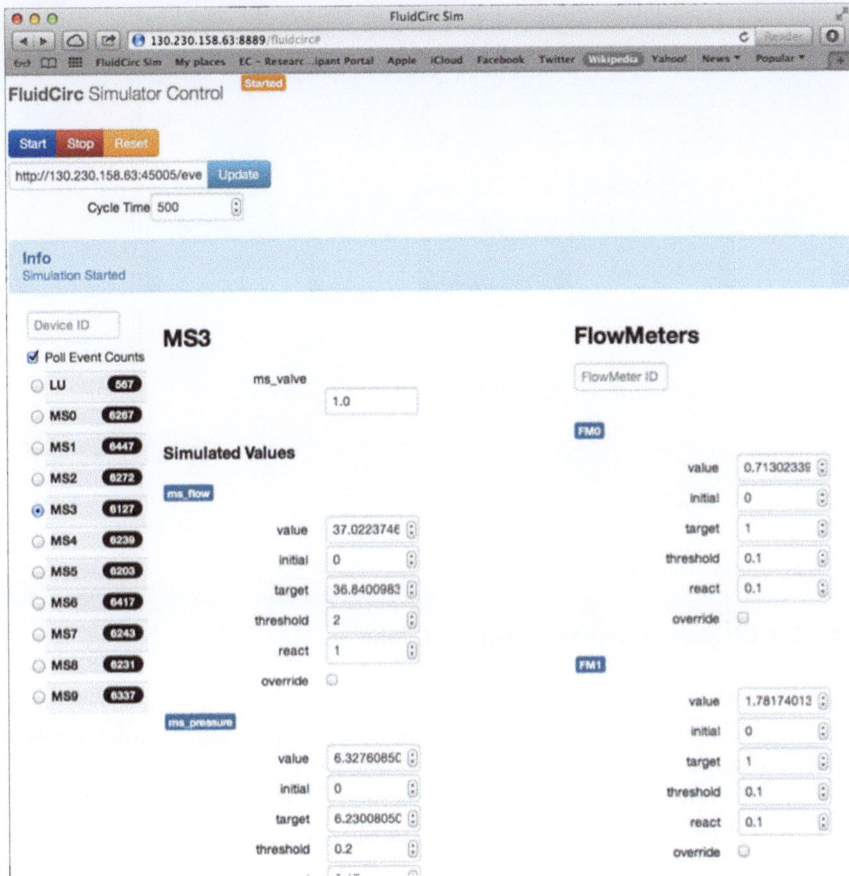

Fig. 8.10 Simulator web interface of an oil lubrication system for paper machines (with MS3 selected)

- One Lubrication Unit (LU);
- Ten Measurement Stations (MS), where each MS is composed of 23 flow metres.

The GUI of the oil lubrication simulator allows users to specify where the events should be output. It shows current values of oil flow rate, pressure and temperature. Additionally, it is possible to see how many events were generated and sent out. The average eventing rate for the events going to CEP service can reach more than 4,500 events per minute. This number was used to check how the system will respond when its input is overloaded. In principle the rates can be adjusted by changing the 'react' parametre (e.g. 0.1), which will produce output event if the change, e.g. in flow metre, will be more or equal to 0.1.

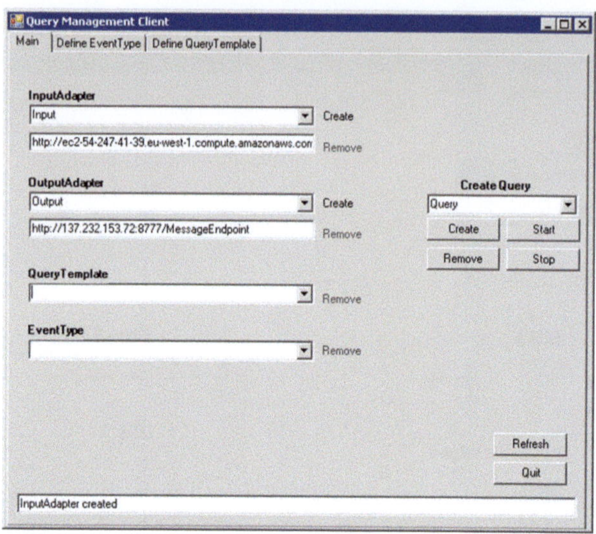

Fig. 8.11 Query management client (main view)

8.4.1.2 CEP Service and Management Client

The CEP Service is configured with 'Query Management Client' (QMC) shown in Figs. 8.11, 8.12 and 8.13. The QMC allows configuring input adapters to define where events can come from, to configure output adapters for specifying output channels for complex events. In principle, the CEP Service can be configured to have several input and adapters output adapters used by a subset of queries. Thus, in principle there can be several applications using CEP Service each working on different subset of input and output adapters with their own CEP queries. Figure 8.12 shows the tab for the event type definition, which is used later as a template for the query. There it is possible to adjust different data types depending on the fields of events to be processed.

Figure 8.13 shows how the query template can be defined depending on event template specified at previous tab. Listing 8.1 shows a query example used for the experiment. The query generates a complex event if it detects a gradual increase 0.1 in flow parametre.

Listing 8.2 shows the SOAP envelope containing the generated complex event. (The gradual increase was detected for 12th flow metre at 8th measurement station, with value '0.306…' at given time). This relatively simple event detecting the gradual increase was used to enable the CEP Service to generate many of those events and to find the limits of the outgoing complex events rate possible for the CEP Service in this particular experimental setup. The content size of the message was 300 bytes.

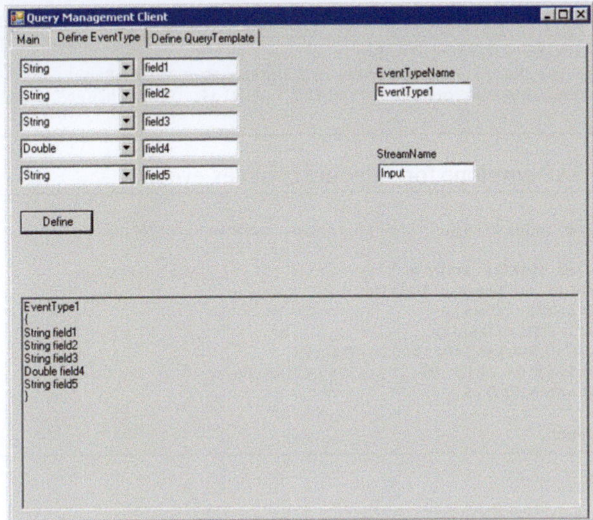

Fig. 8.12 Query management client (event type definition)

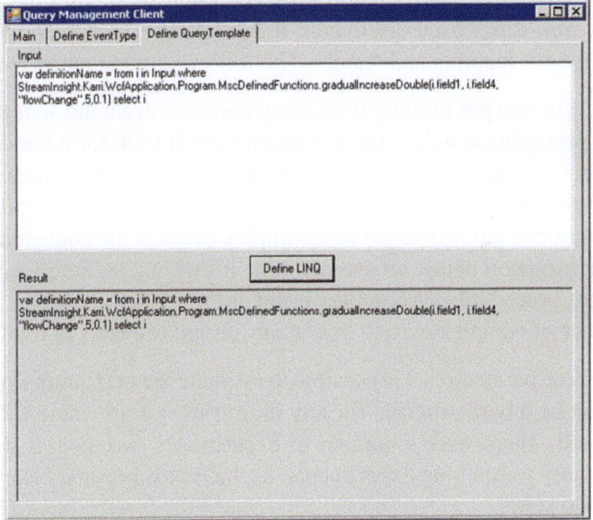

Fig. 8.13 Query management client (query template definition)

Listing 8.1 Test query for evaluating the gradual increase of flow meters

```
var definitionName = from i in Input where
    StreamInsight.Karri.WcfApplication.Program.MscDefinedFunctions.↩
        gradualIncreaseDouble(i.field1, i.field4, "flowChange",5,0.1) select↩
        i
```

Listing 8.2 SOAP envelope for outgoing complex events

```
<SOAp:Envelope xmlns:SOAp="http://schemas.xmlSOAp.org/SOAp/envelope/">
<SOAp:Body>
    <OutputEventSinkQ1 xmlns="http://www.tut.fi/fast/aesop/">
        <Data1>flowChange</Data1>
        <Data2>MS8</Data2>
        <Data3>FM12</Data3>
        <Data4>0.306146240538061</Data4>
        <Data5>08.04.2013 09:21:14.065</Data5>
    </OutputEventSinkQ1>
</SOAp:Body>
</SOAp:Envelope>
```

8.4.1.3 Complex Event Consumer

Complex event consumer was implemented on the Jelastic cloud. It allowed the implementation of a test environment, which can run for a long time without interruption. However, it has been discovered that long experiments are not required to obtain performance features and results. The following parametres were measured:

- Complex event rate per minute: how many complex event are sent per minute.
- Minimum propagation delay: the minimum time it took for a condition (events) from the source to get processed as a complex event at the consumer side.
- Maximum propagation delay: the maximum time it took for a condition (events) from the source to get processed as a complex event at the consumer side.
- Average propagation delay: an average time it took a condition (events) from the source to get processed as a complex event at the consumer side.
- Total number of complex events: how many complex events in total have arrived

Based on these parametres it is possible to estimate the maximum system throughput, which can be a base criterion for any industrial system where CEP is going to be implemented. There were a number of experiments executed. It was found that it is not necessary to run long experiments, as the trends become clear within 5 min after simulation starts.

During longer experiments, it was possible to see that the maximum complex events rate per minute for the given experimental setup could reach 620 complex events per minute. Being overloaded on the input side, the system in general starts to delay sending complex event. In idle conditions, it can take less than 200 ms for the whole loop: (1) event generated at the factory/plant floor, (2) event sent to CEP service, (3) event triggers a complex event and the complex event is delivered and processed by the event consumer.

Table 8.3 A 5-min experiment with incremental grow for flow metres

Measured point	Result
Events sent from oil lubrication simulator	21,157 pcs
Complex event rate per minute	528 complex events per minute
Minimum propagation delay	160 ms
Maximum propagation delay[a]	107,376 ms
Average propagation delay[a]	46,021 ms
Total number of complex events	3,396 pcs

[a] These are not related to the network delays, but are due to overloading the CEP

However, under heavy load, delays start to accumulate, which means that processing of events and generation of complex event is postponed. After running experiments for 1 h the delays were accumulated into 10–12 min taking from a particular event coming from the simulator to trigger corresponding complex event for given experimental setup. It should be noted that while the complex event rate was about 500 events per minute, the CEP service was overloaded with about 4,200 input events coming per minute (Table 8.3). After input events stopped arriving to the CEP service, the output rate for complex events was reaching 620 events per minute. The system was using the time after event flooding stopped to deliver 'old' events. No events were lost.

8.4.2 Conclusions

This experiment has shown that in an event-flooding situation a few minutes are necessary to detect abnormal situations. Under heavy load the time between the condition requiring complex event generation and the actual arrival of the complex event to the dedicated consumer will be growing. This means that after a few minutes these events can be discarded (if it is not desired to log these for later analysis on development of situation at the factory/plant floor).

As a result, it is recommended that operators have a simple mechanism to stop event flooding, but only if the situation starts to develop in this direction (in this case it was just a stop button in oil lubrication simulator as shown in Fig. 8.10). It could also be recommended for a CEP Service to be configurable to define maximal threshold to avoid and/or mitigate event-flooding effects.

In order to estimate thresholds it is required to:

1. Put solution under maximal load.
2. Estimate maximal possible rate for complex event generation (e.g. 620 complex events per minute for given experiment setup).
3. Estimate an average rate between incoming events and complex event generation in extreme case (e.g. how many incoming events in extreme conditions will require a generation of a complex event. For given experimental setup it was $21157/3396 = 6.23$ input events per one outgoing complex event).

4. Ensure required resources for input event processing under extreme case.
5. Define and implement thresholds at each system level (i.e. processing chain) in order to not overload next elements in the event chain. And to define the actions, once the thresholds are reached.

The search of thresholds is required for any application, because no application that cannot be in principle flooded with events exists. Therefore, these steps set the boundaries and evaluate if a CEP solution can be applied in certain cases. Using cloud platforms allows outsourcing of infrastructure and upgrading when necessary. However, it is important to keep in mind that PaaS vendors follow different policies including automatic resource extension depending on run-time needs.

Acknowledgments The authors would like to thank the European Commission for their support, and the partners of the EU FP7 project IMC-AESOP (http://www.imc-aesop.eu) for the fruitful discussions.

References

1. Camp R (2010) An oil flow monitoring system based on web services. Master's thesis. Tampere University of Technology, Tampere
2. Cutler MJ (1996) Paper machine bearing failure. Tappi J 79(2):157–167
3. Day M (1996) Condition monitoring of hydraulic systems. In: Rao B (ed) Handbook of condition monitoring. Elsevier Advanced Technology, Oxford, pp 209–252
4. Holmberg K (2001) Reliability aspects of tribology. Tribol Int 34(12):801–808. http://dx.doi.org/10.1016/S0301-679X(01)00078-0
5. Jestratjew A (2009) Improving availability of industrial monitoring systems through direct database access. In: Kwiecień A, Gaj P, Stera P (eds) Computer networks, communications in computer and information science, vol 39. Springer, Berlin, pp 344–351. doi:10.1007/978-3-642-02671-3_40, http://dx.doi.org/10.1007/978-3-642-02671-3_40
6. Karnouskos S, Colombo AW, Bangemann T, Manninen K, Camp R, Tilly M, Stluka P, Jammes F, Delsing J, Eliasson J (2012) A SOA-based architecture for empowering future collaborative cloud-based industrial automation. In: 38th annual conference of the IEEE industrial electronics society (IECON 2012), Montréal, Canada
7. Rabinowicz E (1981) Lecture presented to the American Society of Lubrication Engineers

Chapter 9
Plant Energy Management

Stamatis Karnouskos, Vladimir Havlena, Eva Jerhotova, Petr Kodet, Marek Sikora, Petr Stluka, Pavel Trnka and Marcel Tilly

Abstract In the IMC-AESOP project, a plant energy management use case was developed to highlight advantages of service orientation, event-driven processing and information models for increased performance, easier configuration, dynamic synchronisation and long-term maintenance of complicated multi-layer solutions, which are deployed nowadays in the continuous process plants. From the application perspective, three scenarios were implemented including advanced control and real-time optimisation of an industrial utility plant, enterprise energy management enabling interactions with the external electricity market, and advanced alarm management utilizing the Complex Event Processing technology.

S. Karnouskos (✉)
SAP, Karlsruhe, Germany
e-mail: stamatis.karnouskos@sap.com

V. Havlena · E. Jerhotova · P. Kodet · M. Sikora · P. Stluka · P. Trnka
Honeywell, Prague, Czech Republic
e-mail: vladimir.havlena@honeywell.com

E. Jerhotova
e-mail: eva.jerhotova@honeywell.com

P. Kodet
e-mail: petr.kodet@honeywell.com

M. Sikora
e-mail: marek.sikora@honeywell.com

P. Stluka
e-mail: petr.stluka@honeywell.com

P. Trnka
e-mail: pavel.trnka@honeywell.com

M. Tilly
Microsoft, Unterschleißheim, Germany
e-mail: marcel.tilly@microsoft.com

A. W. Colombo et al. (eds.), *Industrial Cloud-Based Cyber-Physical Systems*,
DOI: 10.1007/978-3-319-05624-1_9, © Springer International Publishing Switzerland 2014

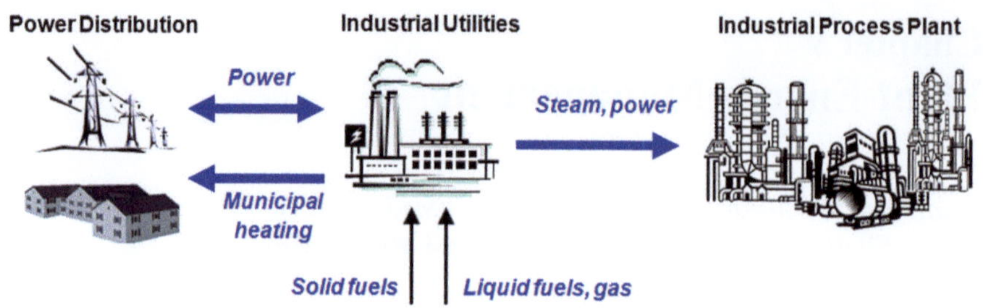

Fig. 9.1 Major flows of energy in industrial utility and process plants

9.1 Introduction

Industrial operating companies have to pay increasing attention to monitoring and optimisation of energy efficiency and carbon emissions. In oil refineries and other enterprises in the petrochemical, chemical, pharmaceutical, or paper-making industry, the utility plant is responsible for the major supply of energy—primarily steam and power—to the process plant (as depicted in Fig. 9.1). The energy can either be generated locally, or purchased from an electricity distribution company. The industrial utility plants may have a contract allowing them to sell excessive amounts of energy back to the electricity grid and take advantage of variable tariffs. Depending on local conditions, they can also serve as a source of heating for residential districts.

Although both industrial utility and process plants are tightly interconnected, their operational and business objectives are different. In the utility plants, the generation of energy is the primary business objective, which is also consistently addressed throughout the facility by adopting hierarchical solutions for closed-loop control and real-time optimisation of individual pieces of equipment (boilers, turbines), their groups (several steam boilers connected to the same header), or the plant as a whole. In contrast to that, the process plants are primarily driven by the objective to produce appropriate mix of products to meet orders coming from the downstream industries.

Energy consumed in the plant and the cost of raw materials are the largest operating costs—and the general desire is to reduce this cost as much as possible, but never in a way that could threaten timely delivery of products. Coupling the industrial utility plants with the envisioned Smart Grid [13] infrastructure in real-time may enable new business opportunities for both sides and enhance energy efficiency. Industrial plants, which have the flexibility to adjust [7] production processes and/or objectives in response to the signals coming from the electricity market, could be seen as an integral part of a larger smart grid ecosystem.

The main objective of the work conducted in the IMC-AESOP project was to highlight advantages of service orientation, event-driven processing and information models for easier configuration, dynamic cross-layer synchronisation and maintenance of complicated multi-layer solutions, which are deployed nowadays in the continuous process plants. From the application perspective, the following three scenarios were implemented:

- energy management of an industrial utility plant enhanced through *cross-layer consistency management*, based on information models,
- *adaptive enterprise energy management* enabling interactions with the external electricity market,
- enhanced operation of processing units through a more effective *alarm management*, driven by the Complex Event Processing technology.

9.2 Cross-Layer Consistency Management

9.2.1 Problem Description

The distributed and networked control of large-scale systems is typically de-signed as multi-layer control architecture, as is the case of industrial utility plants, whose operation is managed by implementing the following application layers:

- *Equipment level*. This basic level is focused on real-time optimisation of individual pieces of equipment—basically the pressure control-related devices like boilers, let down valves and vents, but also other types of more complex equipment including turbo generators or condensing turbines. Advanced process control for boiler modulates fuel feed and air flow to boiler in order to maximize boiler efficiency.
- *Unit level*. Applications at this level deal with the problem of optimal allocation of load between several pieces of equipment running in parallel. This task is usually executed in real-time to ensure fast response to dynamically changing conditions and external requirements. Total steam production is allocated to individual boilers with respect to their efficiency curves to minimize the cost of steam production. The same approach may be applied to multiple turbines.
- *Plant level*. Applications optimize operation of the utility plant over significantly longer periods of time—ranging from hours to days—taking into account multiple possible configurations of the utility plant that can be selected for meeting the energy demand requirements.

The hierarchical approach brings the advantage of a simplified design for complex control strategies, but on the other hand, it complicates the information consistency between individual layers (PID controllers, advanced process controllers, real-time optimizers) under changes in plant topology and other events. Each control layer requires different representation of knowledge, which makes it difficult to guarantee the cross-layer integration, consistency and uniform representation of online and offline process data, topology information and performance models. Within the IMC-AESOP project, these challenges were addressed by implementing a two-level server architecture and OPC-UA [8] information model, which brought the following benefits:

- Information model consistency on all hierarchy levels,
- Cross-layer integration,

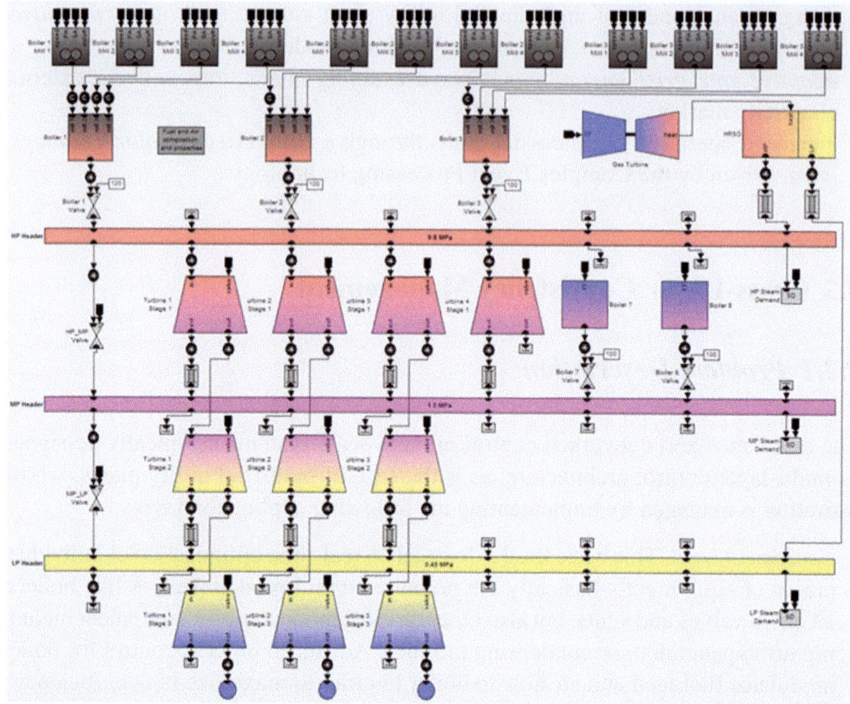

Fig. 9.2 Utility plant model in Matlab Simulink

- Event-driven consistent reconfiguration of all layers,
- Support for flexible on-demand optimisation and what-if analysis,
- Data aggregation from heterogeneous sources.

9.2.2 Information Model

The presented application scenario was primarily focused on the design of information model for a large-scale control system, which was demonstrated on a model of real industrial utility plant (Fig. 9.2). The two-level architecture integrated L1 and L2 information servers (Fig. 9.3) with complementary functions. Raw data collected from the utility plant were aggregated and unified by L1 servers, which are bound and act as a single virtual server containing full OPC-UA information model with data, metadata, and topology information. The aggregating L1 server provides unified access point and event generation to chained L2 servers, which are specialized interfaces mapping L1 information model to a cloud of shared L2 services. The services are higher layers of control hierarchy: advanced process control, real-time optimisation, scheduling, and business planning.

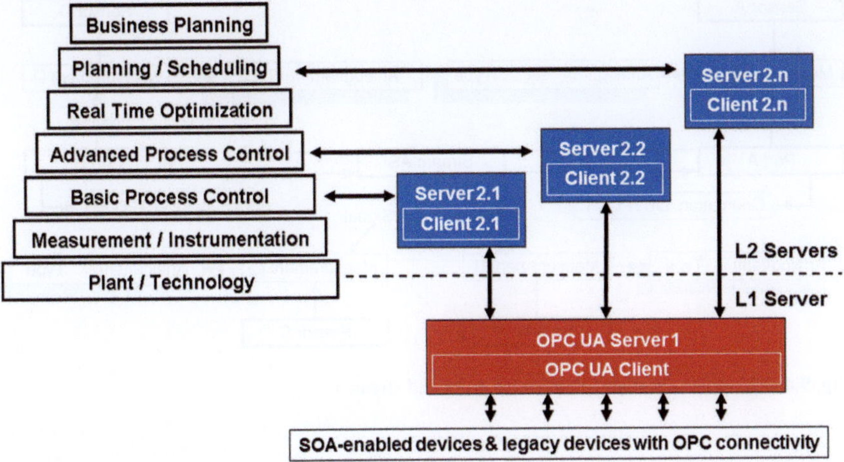

Fig. 9.3 Control hierarchy with layer 1 and layer 2 information servers

The information model was designed to consistently and uniformly represent online and offline process information, which can be used by multiple users with different requirements and functionality. The users can be controllers on a different level of control hierarchy (PID controllers, advanced process controllers, real time optimizers), process operators using their operator screens, alarm management systems, process historian, etc. The information model has two consistent levels. The levels differ in their information details according to their target use.

L1 model is a low-level model containing detailed process information and topology information, which is used by L1 server. The process is represented as a set of *devices*, which have *input* and *output* ports and these ports are interconnected by *streams*. The devices, ports, and streams create basic framework of L1 model [9].

The additional details are usually directly associated with the objects of this basic framework. OPC-UA information model topology is not limited to a tree. It allows "full mesh" topology; however, the backbone of this full mesh structure is usually a tree of hierarchical references, which is used for nodes referencing. The topology information held by L1 model describes interconnections between devices, which always has the following pattern: *Device → Port → Stream → Port → Device*.

Specific object instances can also represent a process value measured by a sensor or a set point for an actuator. This object can be attached directly to a device (measurements or actuators folder) or to a port or stream as illustrated in Fig. 9.4.

L2 model is a high-level model for advanced process control and other higher control layers (APC, RTO, and MES). It contains information necessary for retrieving dynamic or static behaviour models needed by higher control layers. It has dynamic and/or static model of individual devices (examples of model are state space linear model, transfer function matrix, non-linear static model, etc.) and includes description of ports role in controller design (MV—Manipulated Variable, DV—Disturbance

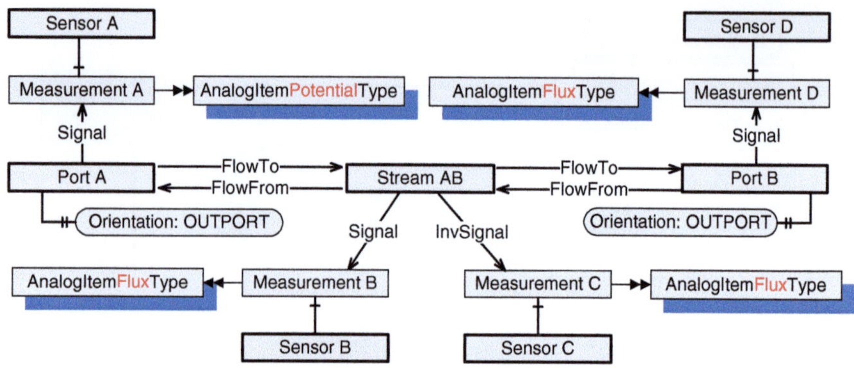

Fig. 9.4 Example of sensors attached to ports and streams

Variable, CV—Controlled Variable, etc.) as well as devices topology description. Generally, the L2 model uses object types of L1 model and extends them with additional functionality wherever needed.

9.2.3 Implementation and Results

The information servers have been implemented in the following way:

- L1 aggregation server holds user-defined OPC-UA address space, including user-defined user type, methods, and references described by standard XML format address space model specified by OPC Foundation.
- It can bind data items to remote servers, working as a client of these servers.
- Upon aggregated data change, it is able to call user defined functions to calculate data from input data item(s) and produce results into output data items. Both input and output data items are represented by OPC-UA variables.
- The chained L2 Server may create subscription to L1 server and monitor data change events on any aggregated or calculated value. Upon that change event each L2 server may perform its own calculations.

Several engineering tools were prototyped to support the whole lifecycle of an OPC-UA information model:

- Information model building tools included the Address Space Model Designer (ASMD), XML editor, and OPC-UA Model Compiler. These tools were used to create the OPC-UA address space, which includes nodes, attributes, and their mutual relationships.
- The data binding tool for binding of the data items inside a server address space to external data sources.

- Information model configuration tool for the chained Level 2 servers where it allows to create an instance of a subsystem and define device, topology, and binding views.

The main impact of the consistent use of information models is on selected engineering aspects associated with the implementation of industrial control solutions. One of the most important is the reduced commissioning effort or reduced number of step tests required for setup of an advanced control solution. For instance, for the utility plant illustrated in Fig. 9.2 the cross-layer consistency service allows to build models for all on/off configurations from models of individual devices. This means that only step testing of individuals devices is required, not the step testing of all possible configurations of the utility plant. Assuming significantly simpler models covering individual devices, the overall effort can be reduced down by tens of percent—this is possible by using algorithms based on the structured model order-reduction [14].

9.3 Adaptive Enterprise Energy Management

9.3.1 Problem Description

The provision of fine-grained information and interaction at enterprise-wide level will have a significant impact on future factories and buildings [12], as well as the associated infrastructure such as EV fleets.

By being able to have fine-grained monitoring and control over the enterprise assets, and access to all information, better planning can be realized, while in parallel efficient strategies can be followed realizing the organisation's objectives such as cost efficiency or sustainable operations. In our case, we investigated the benefits of the enterprise as a whole.

We consider the cogeneration plant (as depicted in Fig. 9.1 and p. xx) as part of the larger picture which includes the infrastructure available, i.e. the company's electric car fleet as well as external energy services such as an energy marketplace. We show that by utilizing the advanced capabilities offered by the energy infrastructure we can realize the vision of more agile enterprises in the future.

This energy management scenario under investigation assumes that the cogeneration utility plant is a source of process steam and electric power for the associated chemical production plant, and additionally, it is also a source of electric power for EV fleet available via the on-site charging stations. The overall goal of the cooperation between the plant and the enterprise energy management system is to cover steam demands and charging needs while following the enterprise's strategic goals. The latter may translate into, e.g. minimizing costs, maximizing profit, increase revenue on energy markets, etc.

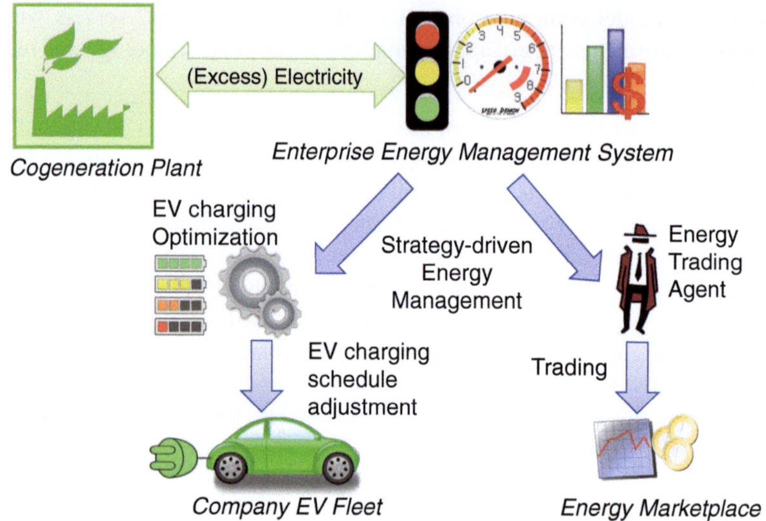

Fig. 9.5 Overview of adaptive energy management with cogeneration plant excess energy availability

9.3.2 Implementation and Results

An overview of the Adaptive Energy Management system is depicted in Fig. 9.5 where we can clearly see the two "tools" available to the enterprise energy management system to achieve its goals, i.e. by deciding to charge electric cars and by using software agents to act on its behalf on energy markets. The decision of one or another tool as well as the potential right mix of them may depend on the actual conditions on the cogeneration plant (e.g. excess energy), needs of the EV fleet, current energy prices on the energy marketplace, potential longer term planning, e.g. storage of energy, etc. Decisions taken may be evaluated also under multiple criteria such as economic benefit, fleet operations, corporate social responsibility goals, etc.

We have looked to two scenarios which represent some key cases:

- *Scenario A.* The energy management scenario assumes that the cogeneration power plant is a source of process steam and electric power for the associated chemical production plant, and additionally, it is also a source of electric power for plant owned EV charging stations. The goal of the cooperation is to cover steam demands and charging needs. In this scenario the decision making is done on the power plant, which regulates how the excess energy can be used and what percentage is tunnelled to the electric car charging and what goes to the energy provider.
- *Scenario B.* In a similar scenario, but this time interacting with a local energy market [2] as envisioned in Smart Grid era, the excess energy not used in plant can be used to charge the Electric Vehicles. Additionally whatever amount of energy is still remaining after the optimal scheduling of cars has been done, can be traded

on the energy market. Similarly, additional energy that might be needed is also acquired from the energy market. The difference with the previous scenario is that it does not assume interaction with a single stakeholder, i.e. the energy provider, and that no adjustments are done on the power plant (hence existing processes remain unchanged). Also the decision making process is now shifted to the orchestrator. This scenario takes advantage of new business opportunities [4], and can be seen as an "add-on" with minimal impact on the power plant operational aspects.

The general workflow is as follows:

- The cogeneration plant is simulated and provides full details on the available excess energy.
- A Decision Support System connects to the simulated plant, and acquires the information. Additionally, it acquires information about the current EV fleet state and plan, as well as info from the energy marketplace.
- After analysis and under the consideration of the enterprise strategies, a decision is taken to (i) store energy by charging the EV fleet, (ii) trade the excess energy on the market, or even a mix among the two that would yield the best benefit, e.g. a financial one.
- Upon request the EV charging optimizer undertakes the task to optimally derive a plan and charge existing (and forthcoming) cars on an optimal schedule that coincides with the available to it excess energy.
- Upon request the Energy Trading Agent connects to the market and places the necessary orders to sell the available energy.
- Information on the results of such actions are communicated back to the cogeneration plant and is depicted in the respective enterprise cockpit.

Some assumptions are made here, as well as some extensions to these actions are possible. For instance it is assumed that the cogeneration plant may rely on an external connection to the grid which takes care of potential imbalances. Additionally deviations are also possible; for instance, if smaller amounts of energy are needed for the EV optimizer due to dynamic events (e.g. larger than expected cars are now requesting charging, etc.) the Energy Trading Agent may issue buy requests to the market to satisfy these needs. Buy requests may also incur also for other reasons, e.g. a fall on the cogeneration plant may result to reduced excess energy being available (than originally predicted) and hence the Energy Trading Agent has "claim back" some of the energy sold which means buying the difference on the market (as one fall-back mechanism if others cannot be realized, e.g. cover the difference with a different EV charging schedule).

The implementation has been realized with the following components and technologies:

- A simulator of the cogeneration plant. This is realized in Matlab/Simulink. Access to the information is provided by an OPC-UA server.
- An "Orchestrator," which assumes the responsibilities of the DSS and orchestrates the integration and decision making. The Orchestrator itself consists of three parts, i.e. an OPC-UA client that connects to the Matlab/Simulink and subscribes to the

events, a Web service client that connects to the EV optimizer cloud service, and a Web service client that connects to the Energy Trading Agent. Additionally, this is the central point for collecting data for future analysis, since it handles the communication with all stakeholders. All of the functionalities related to Web services are developed with the Apache CXF framework that offers RESTful capabilities.

- An EV charging optimizer that optimizes the charging schedule of EVs according to the constraints posed. The EV charging optimizer is realized in Java and runs as a SAP HANA Cloud service. The interfaces it offers are RESTful. The EV charging optimizer considers several dynamic conditions such as production forecast, electricity price, number of expected cars, and tries to find a solution under time constraint (or until it is requested to provide the best solution achieved so far). As our main aim was to demonstrate the easy integration with the IMC-AESOP architecture [5] and external services, we have built upon existing work [10], and extended it for different planning circumstances as well as implemented and deployed in the cloud.

- An Energy Trading Agent and an online marketplace for trading energy at 15-min intervals. All parts here were implemented in Java running as Internet services. We have built upon existing work, i.e. adaptations have been made to connect to an existing Energy Services Platform [6] and to the associated marketplace [2].

The prototype developed as proof of concept has shown that information-driven integration among the various parts of the system can be easily realizable by relying on the IMC-AESOP architecture services [5] and technologies [3]. The usage of Cloud-based services enables the interaction among various stakeholders, and the usage of OPC-UA as well as REST-based Web services acted as enablers for cross-layer information flow and dynamic adjustments.

Although the initial two scenarios presented here validated in the simulation the benefits that could be provided to future businesses, by letting them managing in a more sustainable way their resources, real-world trials under realistic conditions will be needed to further validate the tangible benefits against the cost of implementation, operation and maintenance of such a complex infrastructure. However, the latter should also be assessed from a holistic point of view, for all possibilities they might enable for future enterprises.

9.4 Alarm Management

9.4.1 Problem Description

Due to ever-increasing complexity of production processes and a growing number of collected data points at high sampling frequencies, current process control and monitoring systems are evolving from the synchronous scan-based approach toward the asynchronous event-based paradigm. Complex Event Processing (CEP) represents a scalable and efficient means of handling large amounts of data via event-based communication. Service-Oriented Architecture (SOA) represents another paradigm in control and monitoring systems design. This architectural approach overcomes the problem of great complexity and lack of interoperability and adaptability of current systems. Hence, the demonstrator described in this section aims at overcoming challenges in control system design by exploiting the service architecture and the CEP technology in the design of the alarm system.

The alarm system is a critical part of any process control system. It is designed with the objective to aid the human operators to handle abnormal process situations. A major challenge of current control systems lies in flooding the operator with alarms during process upsets (even if the alarm system is well maintained) [11]. Alarm floods are potentially unsafe, since the operator may overlook important alarms or assess the situation wrongly because of stress and information overload. Alarm floods can be mitigated by the use of advanced alarm management techniques, such as alarm load shedding and state-based alarming, which were prototyped within the IMC-AESOP project.

9.4.2 Alarm Handling Techniques

The following advanced alarm processing functions were implemented within the IMC-AESOP project:

State-based alarming. In certain process states, static alarms can be inadvertently triggered due to normal process changes (e.g. different operating mode or equipment shutdown). In such situations, certain alarms become meaningless or their limits must be set too wide to accommodate the different states. State-based alarming is a dynamic alarm handling method based on switching the alarm system configuration to the settings which correspond to the identified process states. For the different states, new alarms may be enabled, certain alarms may be disabled or their parameters may be altered (such as priority or alarm limit). For the automated switching between configurations, the state detection logic must be reliable and must not chatter [1].

Alarm load shedding. It is a technique that supports operators in prioritizing actions in alarm flooding situations by displaying the most urgent alarms, postponing displaying of less important ones, and filtering out alarms of low priorities. The aim of this method is to keep the alarm rate at a manageable level (ideally one alarm per minute) as applicable. There are two options for triggering this method: manual (by the operator who may select a preconfigured filter) or automatic (based on alarm flood detection). The former approach already occurs in the current practice, while the latter is not yet used.

9.4.3 Architecture

The alarm system architecture developed within the IMC-AESOP project was based on using multiple instances of the CEP engine service and dynamic configuration of the queries executed in this service. Fitting into the context of the SOA-based

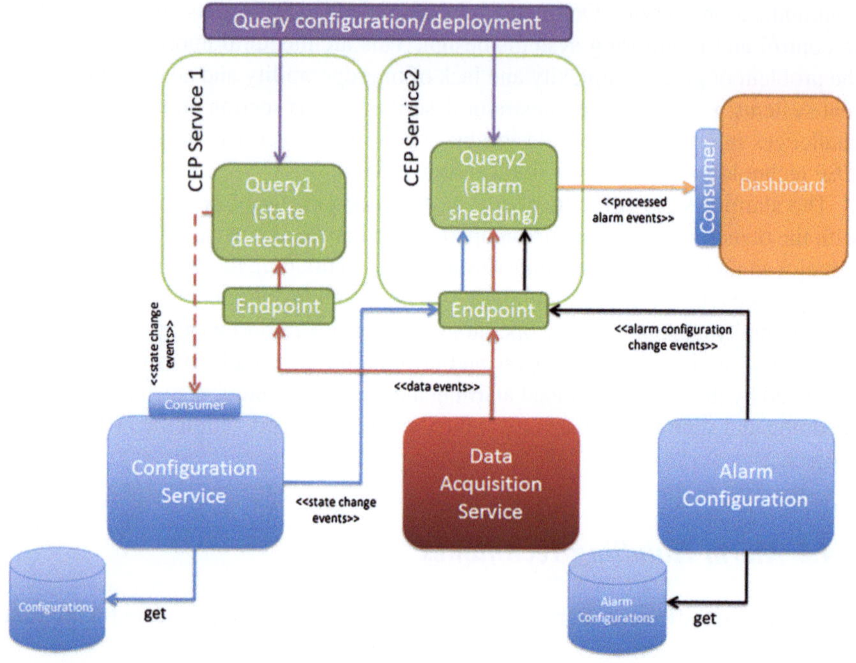

Fig. 9.6 Developed architecture of the alarm system

IMC-AESOP architecture, the following services were implemented to interact with the CEP engine:

- *Data Acquisition Service* delivers data from the process exploiting either the scan-based method (sending values regularly) or the event-based method (delivering values only when there has been any change).
- *Configuration Service* provides information of the current process state (such as startup, normal operation, shutdown, maintenance, fault, off).
- *Alarm Configuration Service* lists all available alarms with their description, properties and settings, and relation to process units and equipment.

The alarm processing functionality is provided by two instances of the CEP engine. The first instance receives current measurements of process variables and based on a dynamically configurable query detects the state of the production process. The detected process state is then sent to the Configuration Service. The other CEP engine instance provides the alarm load shedding functionality defined by another configurable query. This engine also provides the state-based alarming functionality by comparing the incoming measurements against the alarm settings corresponding to the current process state.

All interactions among the services depicted in Fig. 9.6 is strictly event-based. All measurements from the Data Acquisition Service are coming to the CEP engines

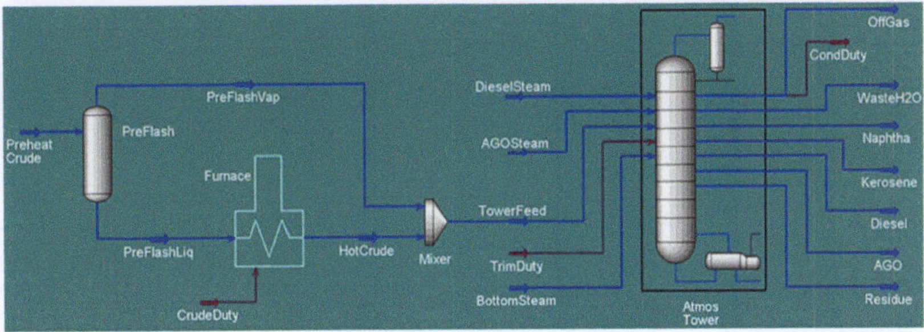

Fig. 9.7 Simulation model of a crude distillation unit

endpoints as events, regardless whether the Data Acquisition Service is implemented as scan-based or event-based. Similarly, the changes of the system state or alarm configuration are also sent as event, which is natively processed by the complex event processing engine.

9.4.4 Simulation Model

A simulation model of a Crude Distillation Unit (CDU) developed in UniSim Design (depicted in Fig. 9.7) was used to measure performance of the new alarm management functions. This model included 131 control system tags sampled with period of 2 s. The data included typical abnormal situations resulting in alarm flood conditions.

   The following states of the CDU were simulated:

- *State 0*: the normal state (the alarm limits design corresponds to light crude oil fed into the column at a medium flow rate),
- *State 1*: light crude oil and a low input flow rate,
- *State 2*: light crude oil and a high input flow rate,
- *State 3*: heavy crude oil and a high input flow rate.

9.4.5 CEP Engine Implementation

The CEP engine based on the Microsoft StreamInsight technology was implemented as a Web service using the standard Web services protocol stack, which makes it well usable in heterogeneous systems. The key point in the implementation is identification of messages/events by the "topic" attribute (see Fig. 9.8), which allows to distinguish between different types of messages. The engine allows the definition of the query (containing the actual instructions for event processing) to be flexible and dynamically configurable via the Management API. The actual implementation of the event processing queries is a standard LINQ standing query as used in Microsoft StreamInsight.

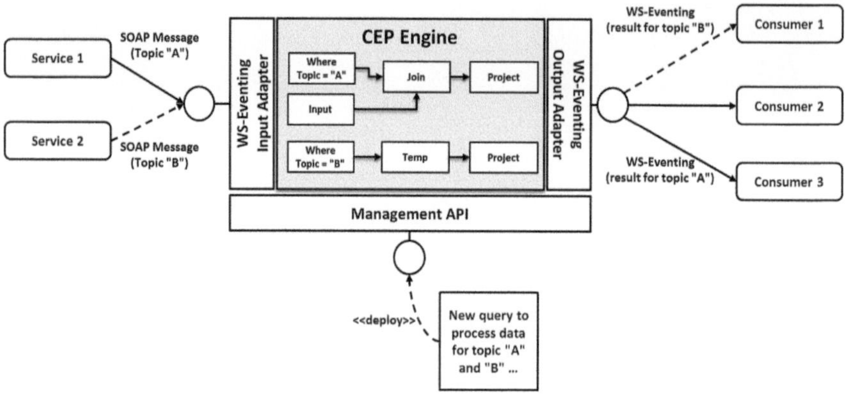

Fig. 9.8 CEP engine as a Web service

Table 9.1 Alarm performance metrics

	Original state (baseline)	State-based alarming	Alarm load shedding	Target
Percent of time in flood (%)	83	31	14	0
Average number of alarms in 10 min	45	8	5	1
Peak number of alarms in 10 min	78	21	13	10
Peak number of alarms in 1 min	65	15	10	2

9.4.6 Results

Performance of the CEP-based alarm management functions was evaluated using the following alarm system performance metrics.

- Percent of time in flood state—The proportion of time that the operator console is flooded with alarms. The rate at which a single operator is overwhelmed by alarm activations (i.e. when the alarm count per 10 min exceeds 10 alarms).
- Average number of alarms in 10 min—The alarm rate that the operator is able to efficiently handle in long term is less than 1 alarm in 10 min.
- Peak number of alarms in 10 min—The maximum rate for the most active 10 min interval within the evaluated time period is 10 alarms.
- Peak Alarm Minute Rate—The target peak minute rate for the most active minute within the evaluated time period. Target = 2/min.

As indicated by Table 9.1, state-based alarming significantly improves all the four evaluated alarm system performance metrics. Alarm load shedding in combination with state-based alarming further improves the results by distributing the alarm load more evenly along the time axis. It also slightly reduces the alarm count, since some the low-priority alarms return to normal due to operator actions addressing other alarms.

9.5 Conclusion

The demonstrators described in this chapter were implemented as a combination of several scenarios highlighting advantages of event-driven processing, service orientation and information modelling for improved cross-layer consistency management, adaptive enterprise energy management and alarm management.

Throughput and service availability measurements showed that cross-layer synchronisation implemented based on OPC-UA does not negatively impact the overall performance of data exchange between individual layers. Moreover, quite significant reductions of engineering efforts—and costs—can be realized by systematic adoption of SOA within industrial process control systems.

Also the consistent use of cloud-based services helped to improve interoperability between various applications as well as interactions among various stakeholders in the energy market scenario. OPC-UA as well as REST-based Web services enabled cross-layer information flow and dynamic adjustments of schedule allowing to respond to market changes in an agile way.

Finally, in the alarm management scenario, CEP engine was harnessed to support implementation of advanced dynamic alarm handling methods, such as state-based alarming and alarm load shedding. The performance measurement was done primarily on the basis of alarm performance metrics whose values indicated significant reduction of alarm flood condition (by 50–70 %) and the peak alarm rates.

Acknowledgments The authors would like to thank the European Commission for their support, and the partners of the EU FP7 project IMC-AESOP (www.imc-aesop.eu) for the fruitful discussions. We would like also to explicitly thank Ji Hu, Mario Graf, Dejan Ilic, and Per Goncalves Da Silva for their contributions.

References

1. Hollifield B, Habibi E (2007) Alarm management: seven effective methods for optimum performance. Instrumentation, Systems, and Automation Society, Germany
2. Ilic D, Goncalves Da Silva P, Karnouskos S, Griesemer M (2012) An energy market for trading electricity in smart grid neighbourhoods. In: 6th IEEE international conference on digital ecosystem technologies—complex environment engineering (IEEE DEST-CEE), Campione d'Italia, Italy
3. Jammes F, Bony B, Nappey P, Colombo AW, Delsing J, Eliasson J, Kyusakov R, Karnouskos S, Stluka P, Tilly M (2012) Technologies for SOA-based distributed large scale process monitoring

and control systems. In: 38th annual conference of the IEEE industrial electronics society (IECON 2012), Montréal, Canada
4. Karnouskos S (2011) Demand side management via prosumer interactions in a smart city energy marketplace. In: IEEE international conference on innovative smart grid technologies (ISGT 2011), Manchester, UK
5. Karnouskos S, Colombo AW, Bangemann T, Manninen K, Camp R, Tilly M, Stluka P, Jammes F, Delsing J, Eliasson J (2012a) A SOA-based architecture for empowering future collaborative cloud-based industrial automation. In: 38th annual conference of the IEEE industrial electronics society (IECON 2012), Montréal, Canada
6. Karnouskos S, Goncalves Da Silva P, Ilic D (2012b) Energy services for the smart grid city. In: 6th IEEE international conference on digital ecosystem technologies—complex environment engineering (IEEE DEST-CEE), Campione d'Italia, Italy
7. Karnouskos S, Ilic D, Goncalves Da Silva P (2012c) Using flexible energy infrastructures for demand response in a smart grid city. In: The third IEEE PES innovative smart grid technologies (ISGT) Europe, Berlin, Germany
8. Mahnke W, Leitner SH, Damm M (2009) OPC unified architecture. Springer, Heidelberg. ISBN 978-3-540-68899-0
9. OPC Foundation (2011) OPC UA specification part 8—data access (RC 1.02)
10. Ramezani M, Graf M, Vogt H (2011) A simulation environment for smart charging of electric vehicles using a multi-objective evolutionary algorithm. In: First international conference on information and communication on technology for the fight against global warming (ICT-GLOW 2011), Toulouse, August 30–31. Lecture notes in computer science, vol 6868. Springer, Berlin, pp 56–63. doi:10.1007/978-3-642-23447-7_6
11. Rothenberg D (2009) Alarm management for process control: a best-practice guide for design, implementation, and use of industrial alarm systems. Momentum Press, New York
12. Sauter T, Soucek S, Kastner W, Dietrich D (2011) The evolution of factory and building automation. Ind Electron Mag IEEE 5(3):35–48. doi:10.1109/MIE.2011.942175
13. SmartGrids ETP (2012) SmartGrids SRA 2035—strategic research agenda. Technical report, SmartGrids european technology platform, European commission. http://www.smartgrids.eu/documents/sra2035.pdf
14. Trnka P, Sturk C, Sandberg H, Havlena V, Rehor J (2013) Structured model order reduction of parallel models in feedback. IEEE Trans Control Syst Technol 21(3):739–753

Chapter 10
Building System of Systems with SOA Technology: A Smart House Use Case

Jerker Delsing, Jens Eliasson, Jonas Gustafsson, Rumen Kyusakov, Andrey Kruglyak, Stuart McLeod, Robert Harrison, Armando W. Colombo and J. Marco Mendes

Abstract The IMC-AESOP architecture has been used to implemente a smart house demonstration. Six different systems has been integrated with local (802.11, 802.15.4) and global (telecom) communication. The six systems integrated are: Car arrival detection system, Garage door opening system, House security system, External house lightning system, External electrical outlet system, House energy control system. The SOA technologies used are CoAP and EXI using SenML to encode the services. Engineering tools have been used to simulate the usage scenario and provide prediction of system behaviour.

J. Delsing (✉) · J. Eliasson · J. Gustafsson · R. Kyusakov · A. Kruglyak
Luleå University of Technology, Luleå, Sweden
e-mail: jerker.delsing@ltu.se

J. Eliasson
e-mail: jens.eliasson@ltu.se

J. Gustafsson
e-mail: jonas.gustafsson@ltu.se

S. McLeod · R. Harrison
University of Warwick, Coventry, UK
e-mail: c.s.mcleod@warwick.ac.uk

R. Harrison
e-mail: robert.harrison@warwick.ac.uk

A. W. Colombo
Schneider Electric, Marktheidenfeld, Germany
e-mail: armando.colombo@schneider-electric.com

A. W. Colombo
University of Applied Sciences Emden/Leer, Emden, Germany
e-mail: awcolombo@technik-emden.de

J. M. Mendes
Schneider Electric, Marktheidenfeld, Germany
e-mail: marco.mendes@schneider-electric.com

A. W. Colombo et al. (eds.), *Industrial Cloud-Based Cyber-Physical Systems*,
DOI: 10.1007/978-3-319-05624-1_10, © Springer International Publishing Switzerland 2014

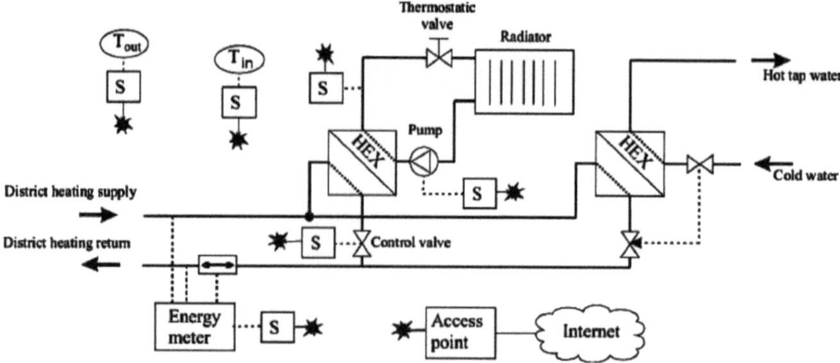

Fig. 10.1 Schematic indication of SOA enabled devices in a district heating substation

10.1 Introduction

Today's technology serves well to build single purpose systems. Lets exemplify this with what's found in a single family house, e.g. a district heating supplied space and tap water heating system, lightning system, security system, tap water distribution system, ventilation, etc. Each of these systems work well on their own with no interaction to other systems. Its not far-fetched that the interaction of these systems can bring added value with respect to energy usage, house security, house owner convenience, etc.

Future houses have to be adaptive to inhabitants and their customs and manners. Thus, houses have to be capable of autonomously providing services like energy usage optimisation, arrival detection and security. Also house hospitality to the house owners and their guests will be of interest. Current technologies can create such solutions but to the expenses of hardware and engineering time to design, build, engineer, deploy and operate bridging technologies for data and information exchange and technology for the cross system integrated services.

The use of SOA enables each individual component in all individual systems to interact [9], which gives users a new way of creating the individual system functionality by engineering new services from primary device supplied services. As such, systems of systems can be created which offer multiple advantages; however they bring also several challenges with them [2, 12]. As an example the necessary control functionality of a district heating substation can be created out of generic services provided by devices like temperature sensors, control valves, energy meters, and pumps. The involved devices and services are indicated in Fig. 10.1. Such system has been previously described [6].

By SOA-enabling each of the individual systems and their components present in a house new system functionality can be created. A use case will be described which has been implemented and tested in a single family house in northern Sweden.

Table 10.1 Services defined and implemented in devices and system of the use case

Name	Usage
Sensory data acquisition	Reads data from temperature sensors, energy meters, valves, power outlets etc
Actuator output	Used to control valves, pumps, power outlets, etc
Historian	Used for logging events and for visualisation purposed
Protocol gateway	Translates between CoAP and HTTP
Time synchronisation	For time stamping of data
Filtering	For filtering sensor values
Monitoring	For monitoring of sensors and control services
Graphic representation	For visualisation of data on the web and user control of actuators

10.2 Use Case Scenario

Using a single family house we will describe and demonstrate the feasibility of the SOA technology to integrate and provide services for energy optimisation, security and even hospitality. For our single family house in northern Sweden, the following separate systems have been SOA enabled.

- Car arrival detection system
- Garage door opening system
- House security system
- External house lightning system
- External electrical outlet system
- House energy control system.

Each of these systems does implement one or many services as provided through the IMC-AESOP architecture [10, 11] using some of the key emerging technologies [7]. Each system is composed of a number of sensors, actuators, and services. The following services are currently defined and used in the demonstrator as shown in Table 10.1.

From these generic services, composed services have been built providing the high-level functionalities as given in Table 10.2.

Apart from the services involved in the demonstration, the event-based SOA system performs closed loop control to control house space heating and tap water heating. All events generated are sent to a Historian service and stored in a MySQL database. Two different web pages are used for data visualisation and user control.

The demonstration has been built up around a single family house. The communication networks used are: local wireless communication between sensors and actuator over 802.15.4 radio connected through an edge router to Internet. Through Internet a cell phone could then be addressed over the present telecom network (GPRS/UMTS/EDGE/LTE) enabling message pushing to, e.g. the house owners cell phone. The local and global communication is depicted in Fig. 10.2.

Table 10.2 High-level functionalities created from generic and composed services in the smart house

Name	Usage
Car arrival detection	Event from road sensor
Car identification	Read of car ID and checking to security system data base
Secure code request	Push of PIN code request to owners cell phone. Owner cell phone number read from security system data base
Opening of garage door	Actuation event created on car id and owner correct PIN code and house energy optima with car parked inside
Enabling out door electric outlet	Actuation event created on car id and owner correct PIN code and house energy optima with car not parked inside
Turning on out door lamp	Hospitality event based on car arrival detection
Secure code request to owners cell phone	Security event based on car arrival detection
Energy optimisation	Car parking in door or out door based on out door temperature level as read from district heating system
Welcome home message to owners cell phone	Pushed to owner cell phone based on car id correct as compared to security data base and correct owner PIN code
Car parking info to owners cell phone	Pushed to owners cell phone based on correct car id and correct owner PIN code and car parking position as determined by energy optimisation

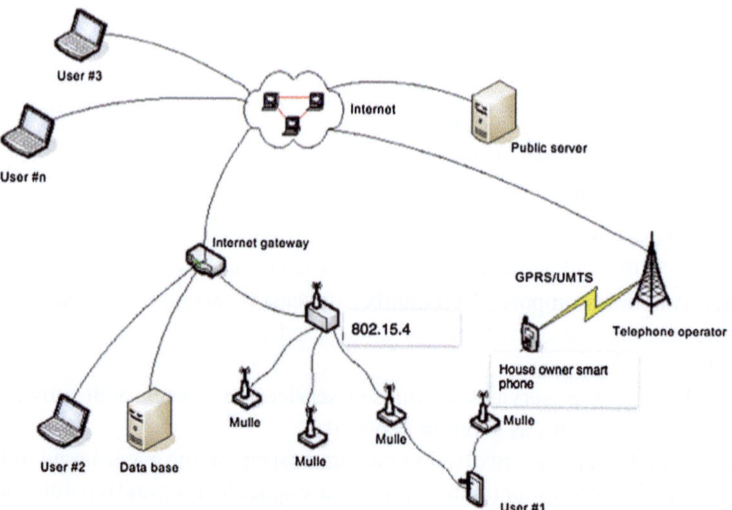

Fig. 10.2 Local and global communication used for the demonstration

For the demonstration, the following technologies were identified as feasible: IPv6, CoAP, EXI and SenML [13]. IPv6 gives global connectivity, CoAP enables SOA functionality and compliance with lightweight devices, EXI compresses XML code service description thus reducing data transfer time and finally SensML provides an established way of describing sensor services. In Fig. 10.3, the used technologies are depicted, all which were implemented on the Mulle platform [8].

Application				
Pilot A Service def	Pilot B Service def	Pilot C Service def	Pilot D Service def	Pilot E Service def
Pilot A SensML	Pilot B JSON def	Pilot C XML def	Pilot D JSON def	Pilot E XML def
XML/JSON- Semantics				
Compression/EXI				
CoAP	DDS	DPWS	PnP	OPC-UA
UDP	HTTP 1.1			
	TCP			
IPv4/IPv6/IP multicast				

Fig. 10.3 SOA protocols being used for the smart house demonstration

Fig. 10.4 Mulle IoT platform (*red ring*) integrated to the space heating control valve of the district heating substation

Mulle is a small (2 cm^2) lightweight IoT platform capable of integrating into most devices. In the demonstration, Mulle was integrated to all necessary devices of the systems used. In Fig. 10.4, the integration of Mulle into the space heating control valve of the district heating substation is shown. The Mulle platform is also the base for road-surface sensors, as described in [4].

With the migration of the necessary devices to the SOA paradigm [3] thus enabling devices to provide services enables composite services to be created in a new manner.

Fig. 10.5 Car being detected by iRoad unit

The composite services for the demonstration, given in Table 10.2, were created in just a few hours by a programmer. The capturing of demonstration scenario scripts was made possible using some of the IMC-AESOP engineering tools. Thus, development and simulation of the demonstration was made enabling the prior understanding and test of system behaviour.

10.3 Demonstration

The demonstration implements the following logic. See also the sequence of Figs. 10.5, 10.6, 10.7.

A car is arriving to the house. The road sensors, developed in the iRoad project [1], detect the arrival and create an event, which is subscribed to by the house security system. This triggers a security control event in the security system. That security event has subevents:

1. Check car identity
2. Push PIN code request to owners cell phone. Car ID and Cell phone number is requested from the security data base service.

In parallel, the car arrival triggers an event in the house energy system. An optimisation algorithm determines from the house energy view point if the car should be parked in the garage or outside and having the car heater outlet enabled. The security event and the energy optimisation event create a decision whether to open the garage or enable the outdoor electric car heating outlet. This triggers the action event of the

Fig. 10.6 House owner is greeted welcome home by the house through the cell phone (*left side*). The house makes the energy optimisation decision and informs the house owner where to park by pushing information to the house owner cell phone

Fig. 10.7 Car parking in the garage upon decision by the house energy optimisation to make use of excess engine heat by opening the garage door and thus allowing the car to be parked in the garage

garage door opening or the enabling of the outdoor outlet, which in turn triggers the owner information event of pushing correct parking information to the house owners cell phone.

The engineering tools developed in IMC-AESOP [12, 15] can capture these type of scripts with logic and timing specifications. An example hereof is shown in Fig. 10.8.

In the PDE toolkit, the systems are broken down into components categorised as actuators, virtual, process, sensors and non-control components. An actuator contains the information of geometry, kinematic and logic, a sensor contains information of geometry and state, virtual components have state behaviour but no geometry or

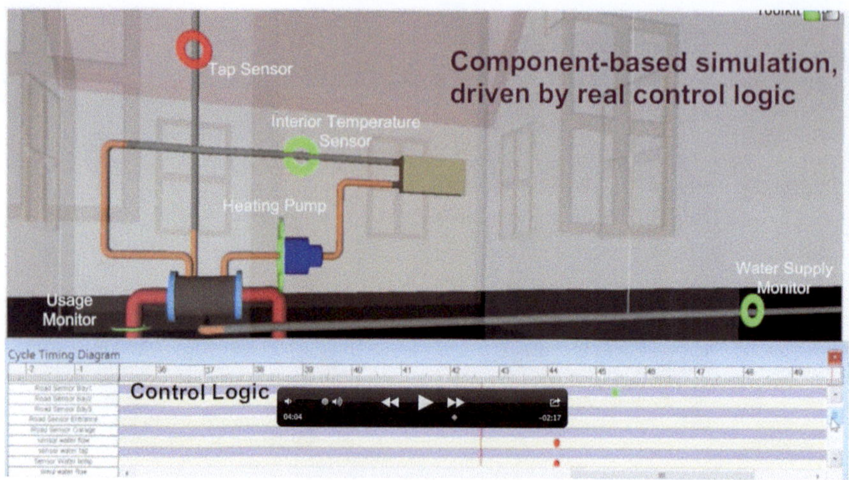

Fig. 10.8 Scene of the demonstration taken from scenario simulation and script capturing tool

physical movement while a non-control only contain geometry information. Process components are very similar to virtual, but are used to describe high-level logic. The logic information of a component is described as a State Transition Diagram (STD). By representing the components using STD interaction between them can be achieved in a structured, readily understood and open way.

A system is built by combining the components geometry and behaviour. The geometric model is defined by linking the geometry using predefined link points associated with a component in a hierarchical manner so that the movement of a parent object is followed by all of its children. Interlocks between states are added to the components to stop unwanted behaviour and collisions during operation, process logic is added to this behaviour for automatic operation.

Figure 10.9 shows a simple example system, the garage port will have state transition diagram as shown. The STDs are defined for all the components and these state diagrams are interlocked to ensure correct. The overall behaviour may be simulated by creating scenarios of use (described using STDs) and using the PDE Orchestrator to execute the logic to validate system behaviour visually. The PDE tools may also be connected to the "real" system to display the actual operation of the "live" systems.

Further evaluation of the control behaviour was achieved using the Continuum tools [14] from the University of Applied Sciences Emden/Leer and Schneider Electric. The Continuum Development Studio (Fig. 10.10) was used to create, analyse and validate using discrete simulation of Petri net-based orchestration topologies (implementing the same component based breakdown and interlocking used in the PDE). The built-in orchestration engine allows to coordinate and synchronise services according to the workflow described by a Petri net model that formalizes the system behaviour. Continuum was also used to ensure correct control operation,

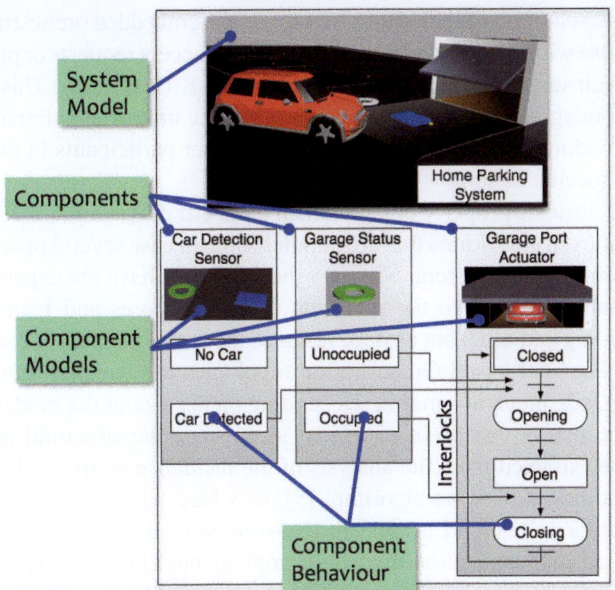

Fig. 10.9 Example component based system described using State Transition Diagrams (STD)

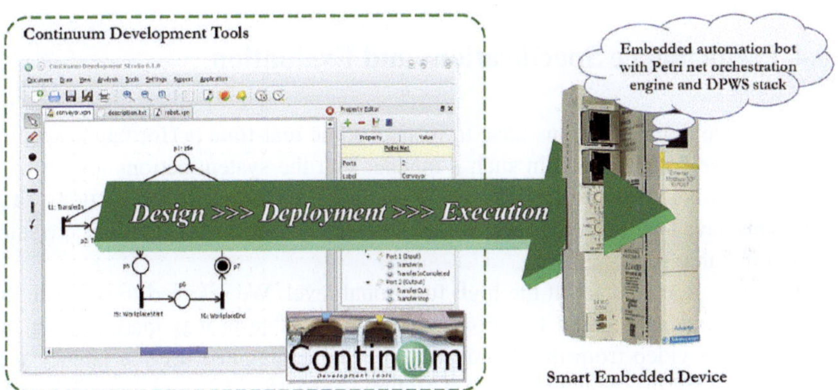

Fig. 10.10 Continuum Development Studio can be used to design, deploy and execute Petri net-based service orchestration specifications

guaranteeing that no deadlocks and livelocks exist, among other structural and behavioural system's specifications.

A similar orchestration engine to the one used in the Continuum Tool for PC can be compiled for embedded devices (such as PLC) with TCP/IP stack support. This embedded orchestration engine can interpret files deployed by the Continuum tools and execute the orchestration locally without the use of the PC where the

Continuum Development Studio runs. Moreover, the embedded orchestration engine can synchronise with other orchestration engines and accept requests or provide status from other Web service endpoints such as MES and ERP systems. This enables the possibility to integrate the system in the "Web service industrial automation cloud". therefore providing the implemented services to other participants in the cloud that may require specific actions.

Due to the intrinsic property of parallelism that Petri nets exhibit, their processing would benefit from algorithms that are parallelisable in case several processing units are available in the host system. Not only the CPU may have the capability of data and task parallelism (due to the inclusion of several cores and features such as Hyper-Threading and SSE) but also for instance the use of Graphics Processing Unit (GPU). The Petri net-based Orchestration topologies are mapped into a matricial formalisation, e.g. incidence matrix. Depending on the size of the modelled system, this incidence matrix can be of enormous size. Since the structural properties of the model are extracted from the analysis of the incidence matrix and this analysis is based on a diagonalisation algorithm [5], it is here where the parallelisation is desired, in order to respond to deal in real-time with the size aspect. Moreover, the parallelised diagonalisation algorithm implemented in the GPU can now run in a concurrently manner to the orchestrated process allowing the support of the process under orchestration by valuable information about structure and behaviour of the system.

10.4 Performance Specifications and Evaluation

The prototype demonstrations have to respect some real-time performance specifications. These were defined in such a manner that the system actions are all time in line with time constants of the physical system. For the human interactions, time constraints have been selected for peoples' convenience. The specifications are summarised in Table 10.3.

These requirements are at the high functional level. We selected to evaluate the specifications at aggregated functional level. A set of tests was made during the recording of a video from the demonstration. The aggregated service latency was in all cases <250 ms, which shows that all timing specifications are met. Thus indicating that the SOA overhead imposed by the technology and the local or global communication constraints does not impair real-time performance at the level of 0.5 s or lower; thanks to the use of the EXI message compression technology.

10.5 Conclusion

The performed real-world demonstrators show that the use of SOA can enable new types of systems, where larger high-end systems can communicate with resource-constrained devices such as sensors and actuators. The demonstration integrates six

Table 10.3 High-level functionality/service real time specifications

Service/functionality	Response time (s)
Car arrival detection	1
Car identification	0.5
Secure code request	0.5
Opening of garage door	0.5
Enabling out door electric outlet	5
Turning on out door lamp	1
Secure code request to owners cell phone	0.5
Energy optimisation	0.5
Welcome home message to owners cell phone	0.5
Car parking info to owners cell phone	0.5
Other functionality	>1

different systems over both global and local wireless communication channels. The performance impact, i.e. overhead, by the use of SOA can be mitigated using efficient data representation and compression schemas. The EXI standard allows verbose, text-based messages to be transferred in a compressed binary form having very little overhead. Thus local to global system services can meet real-time requirements in the range of 0.5 s and slower.

Acknowledgments The authors would like to thank the European Commission for their support, and the partners of the EU FP7 project IMC-AESOP (www.imc-aesop.eu) for the fruitful discussions.

References

1. Birk W, Eliasson J, Lindgren P, Osipov E, Riliskis L (2010) Road surface networks technology enablers for enhanced its. In: Vehicular networking conference (VNC), 2010 IEEE, pp 152–159. doi:10.1109/VNC.2010.5698240
2. Colombo A, Karnouskos S, Bangemann T (2013) A system of systems view on collaborative industrial automation. In: IEEE international conference on industrial technology (ICIT 2013), pp 1968–1975. doi:10.1109/ICIT.2013.6505980
3. Delsing J, Eliasson J, Kyusakov R, Colombo AW, Jammes F, Nessaether J, Karnouskos S, Diedrich C (2011) A migration approach towards a SOA-based next generation process control and monitoring. In: 37th annual conference of the IEEE industrial electronics society (IECON 2011), Melbourne, Australia
4. Eliasson J, Birk W (2009) Towards road surface monitoring: experiments and technical challenges. In: Control applications, (CCA) intelligent control, (ISIC), 2009 IEEE, pp 655–659. doi:10.1109/CCA.2009.5281022
5. Feldmann K, Colombo A (1998) Material flow and control sequence specification of flexible production systems using coloured petri nets. Int J Adv Manuf Technol 14(10):760–774. doi: 10.1007/BF01438228

6. Gustafsson J, Delsing J, van Deventer J (2010) Improved district heating substation efficiency with a new control strategy. Appl Energy 87(6):1996–2004. doi:10.1016/j.apenergy. 2009.12.015

7. Jammes F, Bony B, Nappey P, Colombo AW, Delsing J, Eliasson J, Kyusakov R, Karnouskos S, Stluka P, Tilly M (2012) Technologies for SOA-based distributed large scale process monitoring and control systems. In: 38th annual conference of the IEEE industrial electronics society (IECON 2012), Montréal, Canada

8. Johansson J, Völker M, Eliasson J, Östmark Å, Lindgren P, Delsing J (2004) MULLE: a minimal sensor networking device—implementation and manufacturing challenges. In: Proceedings of IMAPS Nordic, IMAPS

9. Karnouskos S, Colombo AW (2011) Architecting the next generation of service-based SCADA/DCS system of systems. In: 37th annual conference of the IEEE industrial electronics society (IECON 2011), Melbourne, Australia

10. Karnouskos S, Colombo AW, Jammes F, Delsing J, Bangemann T (2010) Towards an architecture for service-oriented process monitoring and control. In: 36th annual conference of the IEEE industrial electronics society (IECON 2010), Phoenix, AZ

11. Karnouskos S, Colombo AW, Bangemann T, Manninen K, Camp R, Tilly M, Stluka P, Jammes F, Delsing J, Eliasson J (2012) A SOA-based architecture for empowering future collaborative cloud-based industrial automation. In: 38th annual conference of the IEEE industrial electronics society (IECON 2012), Montréal, Canada

12. Kaur N, McLeod C, Jain A, Harrison R, Ahmad B, Colombo A, Delsing J (2013) Design and simulation of a SOA-based system of systems for automation in the residential sector. In: IEEE international conference on industrial technology (ICIT 2013), pp 1976–1981. doi:10.1109/ICIT.2013.6505981

13. Kyusakov R, Eliasson J, Delsing J, van Deventer J, Gustafsson J (2013) Integration of wireless sensor and actuator nodes with it infrastructure using service-oriented architecture. IEEE Trans Industr Inf 9(1):43–51. doi:10.1109/TII.2012.2198655

14. Mendes J, Bepperling A, Pinto J, Leitao P, Restivo F, Colombo A (2009) Software methodologies for the engineering of service-oriented industrial automation: the continuum project. In: IEEE 33rd annual international computer software and applications conference, 2009. COMPSAC '09, vol 1, pp 452–459. doi:10.1109/COMPSAC.2009.66

15. Nagorny K, Colombo A, Freijo E, Delsing J (2013) An engineering approach for industrial SOA-based systems of systems. In: IEEE international conference on industrial technology (ICIT 2013), pp 1956–1961. doi:10.1109/ICIT.2013.6505978

Chapter 11
Trends and Challenges for Cloud-Based Industrial Cyber-Physical Systems

Stamatis Karnouskos, Armando W. Colombo and Thomas Bangemann

Abstract The domain of industrial systems is increasingly changing by adopting emerging Internet-based concepts, technologies, tools and methodologies. The rapid advances in computational power coupled with the benefits of the cloud and its services, has the potential to give rise to a new generation of service-based industrial systems whose functionalities reside on-device and in-cloud. Their realisation brings new opportunities, as well as additional challenges. The latter need to be adequately addressed if the vision of future cloud-based industrial cyber-physical system infrastructures is to become a reality and be productively used.

11.1 Vision and Trends

We move towards an infrastructure that increasingly depends on monitoring of the real world, timely evaluation of data acquired and timely applicability of management (control), for which several new challenges arise. Future factories are expected to be complex System of Systems (SoS) that will empower a new generation of today hardly realisable, or too costly to do so, applications and services [6]. New sophisticated enterprise-wide monitoring and control approaches will be possible

S. Karnouskos (✉)
SAP, Karlsruhe, Germany
e-mail: stamatis.karnouskos@sap.com

A. W. Colombo
Schneider Electric, Marktheidenfeld, Germany
e-mail: armando.colombo@schneider-electric.com

A. W. Colombo
University of Applied Sciences Emden/Leer, Emden, Germany
e-mail: awcolombo@technik-emden.de

T. Bangemann
ifak, Magdeburg, Germany
e-mail: thomas.bangemann@ifak.eu

A. W. Colombo et al. (eds.), *Industrial Cloud-Based Cyber-Physical Systems*,
DOI: 10.1007/978-3-319-05624-1_11, © Springer International Publishing Switzerland 2014

due to the prevalence of Cyber-Physical Systems (CPS) [1, 10]. The different systems will be part of a larger ecosystem, where components can be dynamically added or removed and dynamic discovery enables the on-demand information combination and collaboration [3, 4, 17]. All these are expected to empower the transformation to a digital, adaptive, networked and knowledge-based industry as envisioned for Europe [5, 7].

The emerging approach in industrial environments is to create system intelligence by a large population of intelligent, small, networked, embedded devices at a high level of granularity, as opposed to the traditional approach of focusing intelligence on a few large and monolithic applications [3, 4]. This increased granularity of intelligence distributed among loosely coupled intelligent physical objects facilitates the adaptability and reconfigurability of the system, allowing it to meet business demands not foreseen at the time of design and providing real business benefits [13, 16].

Some of the key trends [10] with significant impact on the industrial systems include:

- *Information Driven Interaction*: Future integration will not be based on the data that can be delivered, but rather on the services and intelligence that each device can deliver to an infrastructure. The Service-Oriented Architecture (SOA) paradigm [2] enables abstraction from the actual underlying hardware and communication-driven interaction and the focus on the information available via services.
- *Distributed Business Processes*: In large-scale sophisticated infrastructures, business processes can be distributed in-network, e.g. in the cloud and on the device. Thus processing of information and local decisions can be done where it makes sense and close at the point of action.
- *Cooperating Objects*: Highly sophisticated networked devices are able to carry out a variety of tasks not in a stand-alone mode as usually done today, but taking into full account dynamic and context-specific information. These "objects" will be able to cooperate, share information, act as part of communities and generally be active elements of a more complex system [15].
- *Cloud Computing and Virtualisation*: Virtualisation addresses many enterprise needs for scalability, more efficient use of resources and lower Total Cost of Ownership (TCO) to name a few. Cloud computing has emerged powered by the widespread adoption of virtualisation, service-oriented architecture and utility computing. IT services are accessed over the Internet and local tools and applications (usually via a web browser) offer the feeling that they were installed locally. However the important paradigm change is that the data are computed in the network but not in a priori known places. Typically, physical infrastructure may not be owned and various business models exist that consider access-oriented payment for usage.
- *Multi-Core Systems and GPU Computing*: Since 2005 we have seen the rapid prevalence of multi-core systems that nowadays start to dominate everyday devices such as smartphones. The general trends are towards chips with tens or even hundreds of cores. Advanced features such as simultaneous multi-threading,

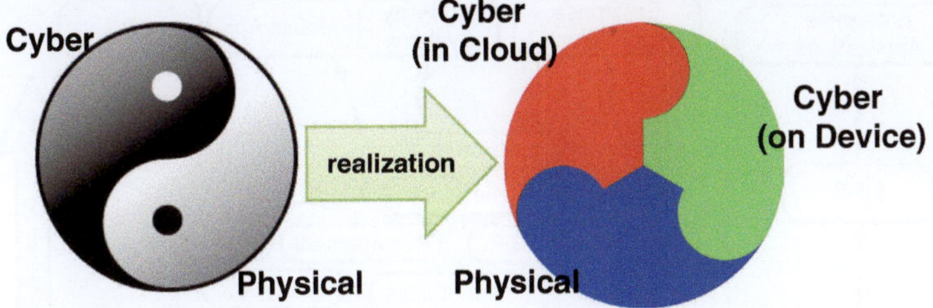

Fig. 11.1 Cloud-based cyber-physical systems

memory-on-chip, etc., promise high performance and a new generation of parallel applications unseen before in embedded systems. Additionally, in the last decade we have seen the emergence of GPU computing where computer graphic cards are taking advantage of their massive floating-point computational power to do stream processing. For certain industrial applications this may mean a performance increase of several orders of magnitude compared with a conventional CPU.

- *SOA-Ready Devices*: Networked embedded systems have become more powerful with respect to computing power, memory and communication; therefore they are starting to be built with the goal to offer their functionality as one or more services for consumption by other devices or services. Due to these advances we slowly witness a paradigm shift where devices can offer more advanced access to their functionality and even host and execute business intelligence, therefore effectively providing the building blocks for expansion of SOA concepts down to their layer. Web services are suitable and capable of running natively on embedded devices, providing an interoperability layer and easy coupling with other components in highly heterogeneous shop-floors [3, 4, 13, 16].

All of the aforementioned trends, heavily impact the next generation of industrial systems per se. Cloud integration and interaction may be a game-changer in the CPS and will take a closer look at how CPS and cloud fuse.

11.2 The Fusion of Cloud and CPS

The first step in the infrastructure evolution was to empower individual devices with Web services, and enable them to participate in a service-based infrastructure. This is achieved by enabling them to (i) expose their functionalities as services and (ii) empower them to discover and call other (web) services to complement their own functionalities [3, 13, 16].

The next step is to take advantage of modern capabilities in software and hardware, such as the cloud and the benefits it offers. As seen in Fig. 11.1, CPS have two key parts integrated in balance, the physical part for interacting with the physical environment (e.g. composed of sensors and actuator constellations), and the cyber part, which is the software part managing and enhancing the hardware capabilities of

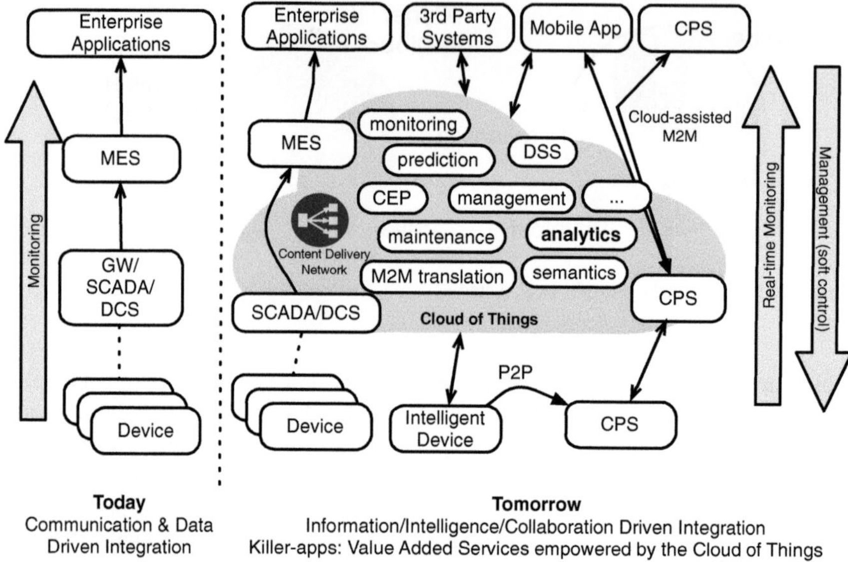

Fig. 11.2 Vision of cloud-based SOA industrial systems

the CPS as well as its interaction with the cyber-world. The prevalence of the cloud and its benefits [11], enables us to expand the cyber-part of the CPS and distribute it on-device and in-cloud. As depicted in Fig. 11.1, CPS now may operate with three key parts constituting and forming their interaction in the physical and virtual world.

The cloud-enabled CPS have profound implications for the design, development and operation of CPS. Although the device-specific part, i.e. the cyber (on-device) and physical part are still expected to work closely together and provide the basic functionalities for the CPS, the in-cloud cyber part may evolve independently. Due to its nature, the in-cloud part will require connectivity of the CPS with the cloud where added-value sophisticated capabilities may reside. On the contrary the on-device cyber-part may consider opportunistic connections to the cloud, but in general should operate autonomously and in-sync with the physical part.

The nature of the functionalities as well as the degree of their dependence on external resources, computational power, operational scenarios, network connectivity, etc., will be the key factor for hosting them on-device or in-cloud. Nevertheless, typical considerations up to now about resource-constrained devices do not hold in general any more, as now the additional power needed from specific functionalities can be outsourced to the cloud and hence the software/hardware needs for these functionalities is no longer required to be on the device itself [11]. The latter enables more flexibility for the design and operation of large industrial CPS infrastructures that act collaboratively, and may achieve more, by better utilising their resources.

As an example of this era, we have to point out clearly that the next-generation SCADA/DCS systems may not have a physical nature but rather rely on federated

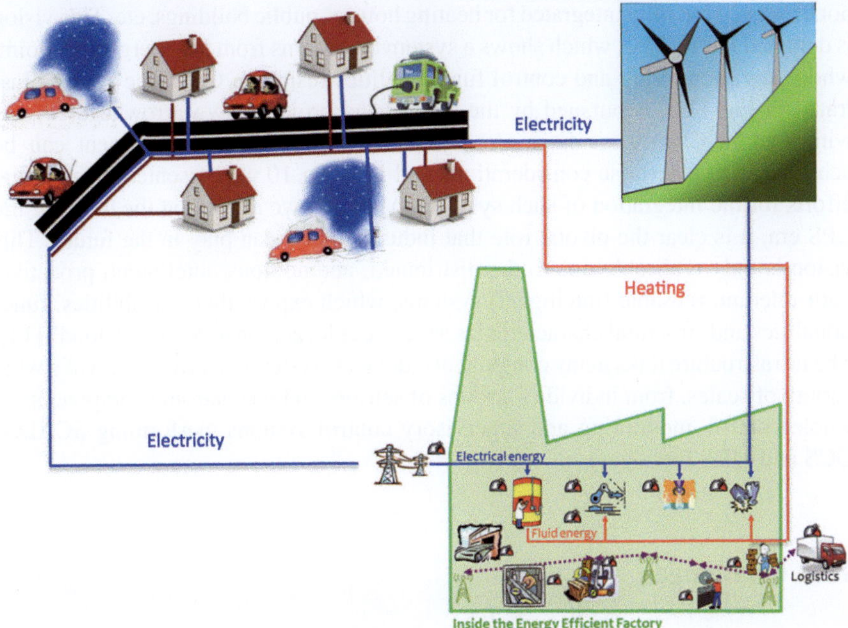

Fig. 11.3 A System of systems view empowered by CPS for the energy domain

actuators and sensors, while their main functionalities reside solely on the cloud [10]. This implies that it might reside only on the cyber or "virtual" world, in the sense that it will comprise multiple real-world devices, on-device and in-network services and collaboration-driven interactions that will compose a distributed highly agile collaborative complex system of systems.

As shown in Fig. 11.2, the fusion of CPS and Cloud constitutes the "Cloud of Things" [11], which flourishes based on services offered to devices and systems, as well as depend on data from devices and intelligence built on the interaction among the physical and cyber (virtual) world. The benefit of utilising the Cloud of Things is that additional capabilities potentially not available at resource constraint devices can now be fully utilised taking advantage of cloud characteristics such as virtualisation, scalability, multi-tenancy, performance, life cycle management, etc. The manufacturer for instance can use such cloud-based services to monitor the status of the deployed appliances, make software upgrades to the firmware of the devices, detect potential failures and notify the user, schedule proactive maintenance, get better insights into the usage of his appliance and enhance the product, etc.

CPS are seen as a key part of critical infrastructures including the energy domain [8]. Future smart cities will integrate multiple such systems in a harmonised way to enable new innovative services for their citizens. Hence, factories will be situated within cities, smart buildings and smart houses will take full advantage of the energy available in the grid, and all forms of energy by-products such as heat will not be wasted but fully integrated for heating houses, public buildings, etc. This vision is depicted in Fig. 11.3, which shows a system of systems from the energy viewpoint, whose key monitoring and control functionalities reside on CPS. The vision illustrated in Fig. 11.3, is pursued by the Arrowhead project (www.arrowhead.eu). As witnessed in Chap. 9 we have already shown how energy management can be achieved with enterprise considerations and in Chap. 10 we presented some initial efforts for the integration of such systems. Although we are still at the dawn of the CPS era, it is clear the pivotal role that industrial CPS can play in the future. This vision is only realisable due to the distributed, autonomous, intelligent, proactive, fault-tolerant, reusable (intelligent) systems, which expose their capabilities, functionalities and structural characteristics as services located in a "Service Cloud" [14]. The infrastructure links many components (devices, systems, services, etc.) of a wide variety of scales, from individual groups of sensors and mechatronic components to whole control, monitoring and supervisory control systems, performing SCADA, DCS and MES functions.

11.3 Challenges

For the new infrastructure to materialise and become a reality, several challenges need to be adequately addressed. We indicate here some key questions on which more research and experimentation will need to be conducted to assess their impact on future industrial CPS systems, as well as the degree of their fulfilment that is required, especially for the critical infrastructures. We depict here some thoughts for consideration:

- *Management*: Considering the hundreds of thousands of devices only active in an industrial setting, e.g. a factory, or the millions of them in a larger one, e.g. a smart city, new ways of easily managing large-scale and complex systems need to be considered. Dynamic discovery, interaction and exchange of information as well as life cycle management especially over federated systems is challenging.
- *Security, Trust, Resilience, Reliability and Safety*: CPS have a real-world impact and control real-world infrastructures. Failures may result in havoc with escalated effects that may impact safety. To what extent such systems can be designed with security, trust and safety in mind, especially when operating as part of a larger ecosystem is not trivial [8]. The tackling of reliability of CPS ecosystems as well as that of resilience will be the key factor for their application in critical systems, or otherwise put, to what extent our core critical infrastructure will be vulnerable in the future [9].
- *Real-time Data Collection—Analysis—Decision—Enforcement*: For CPS to excel in their role, real-time collection of data has to be realised, and subsequently its analysis can help take the appropriate business decisions and enforce them. Although CPS up to now had local decision loops, with fusion with the cloud and dependence on external services, the timely interaction aspects need to be revis-

ited. A distributed collaborative approach is called upon here, where parts of the functionalities are hosted where it makes sense (on-device, in-cloud etc.) to guarantee real-time interactions from data-collection, for analysis, decision and enforcement.

- *Cross-layer Collaboration*: CPS and their effectiveness will depend on the collaboration with other CPS and systems via a service-based infrastructure as already analysed. However, such complex collaborations will have various requirements from the technical and business side that will need to be respected, depending on the application scenario. How to effectively empower collaboration via services and tools, including interactions in intra- and cross-domain so that emergent behaviour can flourish in ecosystems of CPS, is not an easy undertaking.

- *Semantic-driven Discovery and Interaction*: Discovering the right services based on functionalities they provide, being able to communicate and exchange interoperable data and build collaborations, is a key enabler for future CPS. However, how this can be realised for multiple domains, dominated by a plethora of heterogeneous (in hardware and software) systems and services, is a grand challenge.

- *Application Development Based on Generic CPS APIs*: CPS APIs reflecting the core functionalities need to be present and offer standardised interactions upon which more complex behaviours and services can be built. This will act as an enabler in the short term until the semantic-driven interaction is fully tackled. Applications and services can then build upon the minimum services offered by the CPS itself as well as its envisioned supporting infrastructure (CoT) and extend them.

- *Migration and Impact of CPS to Existing Approaches*: The introduction of CPS will ignite a rethinking on various levels at the infrastructure itself as well as the processes that depend on it. However, assessing the exact impact on a larger scale system might be challenging and has to be carefully investigated. As CPS will gradually replace legacy approaches, strategies for the migration of legacy systems to CPS ones are needed. To this end, simulators/emulators of systems and behaviours are also needed to assist with the assessment of transitions.

- *Sustainable Management*: Cloud-based CPS bring the promise for more efficient usage of the globally available resources as well as optimisations from various perspectives, e.g. execution, communication, interaction, management, etc. Hence more sustainable strategies for managing infrastructures and businesses may be realised, e.g. energy-driven management [12]. Such efforts should be seen in a greater context, i.e. cross-enterprise, smart city-wide, etc. Tools and approaches that will empower us to integrate such approaches effectively in large-scale CPS are needed.

- *Development and Engineering Tools*: Development and Engineering tools and environments will be a must to ease the CPS ecosystem service creation and orchestration/choreography within complex environments. Cross-platform availability and capability are seen as key aspects for offering sophisticated services. These tools will need to be coupled with appropriate "wizards", debugging capabilities (at local and system-wide level), as well as simulation environments where "what-if" approaches can be realised.

- *Data Life Cycle Management & Sharing*: Being able to acquire the data from the physical and cyber world is the first step. Sharing them in order to built sophisticated services and effectively managing them is a grand challenge. The latter has to be done with consideration of the operational context its requirements for security, privacy, etc., while in parallel enabling their wide availability, e.g. as open data in appropriate forms for other parties to extract information for their processes. Although the specific business needs and requirements have to be satisfied, data from CPS will be a commodity in the years to come, and will be traded as such.
- *Data Science on CPS-Empowered Big Data*: The massive CPS infrastructures envisioned and their fusion with the cloud, will lead to massive amounts of data acquired for the finest details of a process. This "Big Data" can be analysed in the cloud and provide new insights for the industrial processes that may lead to better enterprise operations and identification of optimisations. Data science approaches on the available Big Data is expected to have a wide impact on the way we design and operate CPS infrastructures.

Industrial cyber-physical systems are changing the economy and society [1]. Therefore, in addition to the key challenging aspects raised above, one has to always bear in mind that CPS will have to address the human factor adequately in order to be successful. This puts a spotlight on another set of challenges such as:

- *Education*: Due to the complexity and sophistication of CPS and the domains in which they are applied, a new generation of engineers will have to be educated on a variety of aspects pertaining to several domains. This implies cross-disciplinary skills that successfully fuse application domain-specific knowledge, CPS engineering as well as HCI skills that will need to be continuously maintained (life-long learning). Such programmes should be introduced at universities at graduate and postgraduate levels including specialisation on CPS technologies.
- *Training*: The industrial adoption of this new paradigm represents a revolution that requires advanced skills and extensive training activity. Architects, engineers and operators at first level, will need to be re-educated for dealing with heterogeneous physical and cyber systems, as well as fully understand their capabilities, benefits and challenges they offer. Simulation/Emulation and hands-on experiences are considered pivotal towards tackling this challenge.
- *Thinking Shift*: The benefits can be tremendous in B2B, B2C, B2B2C, etc., and grasping the potential as well as correctly assess the risk associated means that not only new business models should be developed, but increasingly focus on the human role in these as an end-use of a CPS (either directly or via the surrounded infrastructure). Decision-makers, industrial strategists, legislators and policy-makers will have to consider a balanced action for empowering innovation without falling short on privacy, usability, espionage, security and trust.

11.4 Conclusion

We have presented a vision, some major trends that will reshape the way we design, implement and interact in future industrial CPS-dominated environments, especially when it comes to monitoring and management, as well as some key challenges and considerations. The fusion of cyber-physical systems with the cloud is still at a very early stage. However, it has profound implications as it blurs the fabric of cyber (business) and physical worlds. Time-sensitive monitoring, analytics and management will be of key importance for any real-world application. As such, emphasis should be given to the basic parts of such collaborative CPS ecosystems to act as enablers towards vision realisation. The considerations raised here for CPS to be used in industrial applications, are in the same line of thought as the recommendations for action [1] for the successful introduction and widespread adoption of CPS in general. Only then can key industrial visions such as the Industry 4.0 [7] materialise.

Acknowledgments The authors thank the European Commission for their support, and the partners of the EU FP7 project IMC-AESOP (www.imc-aesop.eu) for fruitful discussions.

References

1. acatech (2011) Cyber-physical systems: driving force for innovation in mobility, health, energy and production. Technical report, acatech—National Academy of Science and Engineering. http://www.acatech.de/fileadmin/user_upload/Baumstruktur_nach_Website/Acatech/root/de/Publikationen/Stellungnahmen/acatech_POSITION_CPS_Englisch_WEB.pdf
2. Boyd A, Noller D, Peters P, Salkeld D, Thomasma T, Gifford C, Pike S, Smith A (2008) SOA in manufacturing—guidebook. Technical report, IBM Corporation, MESA International and Capgemini. ftp://public.dhe.ibm.com/software/plm/pdif/MESA_SOAinManufacturingGuidebook.pdf
3. Colombo AW, Karnouskos S (2009) Towards the factory of the future: a service-oriented cross-layer infrastructure. In: ICT shaping the world: a scientific view. European Telecommunications Standards Institute (ETSI), Wiley, New York, pp 65–81
4. Colombo AW, Karnouskos S, Mendes JM (2010) Factory of the future: a service-oriented system of modular, dynamic reconfigurable and collaborative systems. In: Benyoucef L, Grabot B (eds) Artificial intelligence techniques for networked manufacturing enterprises management. Springer, London. ISBN 978-1-84996-118-9
5. European Commission (2004) Manufuture: a vision for 2020. http://www.manufuture.org/documents/manufuture_vision_en%5B1%5D.pdf, report of the high-level group
6. Jamshidi M (ed) (2008) Systems of systems engineering: principle and applications. CRC Press, Boca Raton
7. Kagermann H, Wahlster W, Helbig J (2013) Recommendations for implementing the strategic initiative INDUSTRIE 4.0. Technical report, acatech—National Academy of Science and Engineering. http://www.acatech.de/fileadmin/user_upload/Baumstruktur_nach_Website/Acatech/root/de/Material_fuer_Sonderseiten/Industrie_4.0/Final_report__Industrie_4.0_accessible.pdf
8. Karnouskos S (2011a) Cyber-physical systems in the smartGrid. In: IEEE 9th international conference on industrial informatics (INDIN), Lisbon, Portugal

9. Karnouskos S (2011b) Stuxnet worm impact on industrial cyber-physical system security. In: IECON 2011—37th annual conference on IEEE industrial electronics society, pp 4490–4494. doi:10.1109/IECON.2011.6120048

10. Karnouskos S, Colombo AW (2011) Architecting the next generation of service-based SCADA/DCS system of systems. In: 37th annual conference of the IEEE industrial electronics society (IECON 2011), Melbourne, Australia

11. Karnouskos S, Somlev V (2013) Performance assessment of integration in the cloud of things via web services. In: IEEE international conference on industrial technology (ICIT 2013), Cape Town, South Africa

12. Karnouskos S, Colombo A, Lastra J, Popescu C (2009) Towards the energy efficient future factory. In: 7th IEEE international conference on industrial informatics, INDIN 2009, pp 367–371. doi:10.1109/INDIN.2009.5195832

13. Karnouskos S, Savio D, Spiess P, Guinard D, Trifa V, Baecker O (2010) Real world service interaction with enterprise systems in dynamic manufacturing environments. In: Artificial intelligence techniques for networked manufacturing enterprises management. Springer, London

14. Karnouskos S, Colombo AW, Bangemann T, Manninen K, Camp R, Tilly M, Stluka P, Jammes F, Delsing J, Eliasson J (2012) A SOA-based architecture for empowering future collaborative cloud-based industrial automation. In: 38th annual conference of the IEEE industrial electronics society (IECON 2012), Montréal, Canada

15. Marrón PJ, Karnouskos S, Minder D, Ollero A (eds) (2011) The emerging domain of cooperating objects. Springer, Berlin. http://www.springer.com/engineering/signals/book/978-3-642-16945-8

16. Mendes J, Leitão P, Restivo F, Colombo AW (2009) Service-oriented agents for collaborative industrial automation and production systems. In: Mařík V, Strasser T, Zoitl A (eds) Holonic and multi-agent systems for manufacturing, Lecture Notes in Computer Science, vol 5696, Springer, Berlin, pp 13–24. doi:10.1007/978-3-642-03668-2_2, http://dx.doi.org/10.1007/978-3-642-03668-2_2

17. Sauter T, Soucek S, Kastner W, Dietrich D (2011) The evolution of factory and building automation. Ind Electron Mag IEEE 5(3):35–48. doi:10.1109/MIE.2011.942175

Book Editors

Prof. Dr.-Ing. Armando Walter Colombo joined the Department of Electrotechnic and Industrial Informatics at the University of Applied Sciences Emden/Leer becoming Full Professor of Industrial Informatics and Automation in 2010. He is currently Director of the Institute for Industrial Informatics, Automation and Robotics (I2AR). Prof. Colombo is also Edison L2 Group Senior Expert and Research Program Manager at Schneider Electric Automation GmbH. He received the Doctor degree in Engineering from the University of Erlangen-Nuremberg, Germany, in 1998. From 1999 to 2000 he was Adjunct Professor in the Group of Robotic Systems and CIM, Faculty of Technical Sciences, New University of Lisbon, Portugal. He has extensive experience in managing multi-cultural research teams in multi-regional projects. Prof. Colombo has participated in leading positions in many international projects, e.g. he was co-leader of the RTD-Cluster on Production Automation and Control (PAC) of the EU FP6 NoE 'IPROMS' (www.iproms.org, 2004–2009), technical manager of the EU FP6 STREP RI-MACS (2005–2008), coordinator of the EU FP6 Integrated Project 'SOCRADES' (www.socrades.eu, 2006–2009), with the participation of all major European Stakeholders of the Automation value chain, i.e. Schneider Electric, Siemens, ABB, but also ARM, SAP, Jaguar/Ford, etc., and coordinator of the EU FP7 Integrated Project 'IMC-AESOP' (www.imc-aesop.eu, 2010–2013). His research interests are in the fields of Cyber-Physical Systems (CPS), Service-Oriented Architecture (SOA), collaborative automation, intelligent supervisory control and formal specification of flexible automation systems. Prof. Colombo has more than 200 publications (peer-reviewed) in journals, books and chapters of books and conference proceedings, and is co-author of 23 industrial patent applications (see http://scholar.google.com/citations?user=csLRR18AAAAJ). He is a senior member of the IEEE, member of the Administrative Committee (AdCom) of the IEEE Industrial Electronic Society and member of the Gesellschaft für Informatik e.V. Prof. Colombo is the Schneider Electric representative in ARTEMIS (European Embedded Systems Platform), co-leading the Sub-Program ASP 4 associated to the ARTEMIS Strategic Research Agenda. He served/s as advisor for the definition of the R&D priorities within the last three Framework Programs of the European Commission.

A. W. Colombo et al. (eds.), *Industrial Cloud-Based Cyber-Physical Systems*,   241
DOI: 10.1007/978-3-319-05624-1, © Springer International Publishing Switzerland 2014

Prof. Colombo is listed in the Who's Who in the World /Engineering 99-00/01 and in Outstanding People of the Twentieth Century (Bibliographic Centre Cambridge, UK).

Dr. Thomas Bangemann is Deputy Head of the ifak Institut für Automation und Kommunikation e.V. Magdeburg. Formerly, he headed the departments of IT and Automation as well as of Industrial Communication Systems at ifak. After he finished his scientific studies on monitoring, control and diagnostics of automation systems in 1993 with the doctoral level, he has been working on the subjects of communication systems and their application, application of automation systems, introduction of information technologies to management applications as well as integration of automation systems into SOA-based systems. During the last few years he has been involved in several European and national funded projects, e.g. SOCRADES, AIMES, PROTEUS or IMC-AESOP. He is a member of several working groups within the ZVEI (Manufacturing Execution Systems, Steering Committee for Communication in Automation), VDI/VDE-GMA (Cyber Physical Systems) and PROFIBUS International and he also gives lectures on Process Control at the University of Applied Sciences Magdeburg-Stendal.

Stamatis Karnouskos is with SAP as a Research Expert on Internet of Things. He investigates the added-value of integrating networked embedded devices in enterprise systems. For more than 15 years Stamatis has led efforts in several European Commission and industry funded projects related to industrial automation, smart grids, Internet-based services and architectures, software agents, mobile commerce, security and mobility. Stamatis is actively involved in several consultations at the European Commission and German level dealing with Cyber-Physical Systems, System of Systems, Internet of Things, energy efficiency and SmartGrids. He has co-authored and edited several books, over 150 technical articles, acted as guest editor in IEEE/Elsevier journals, and participates as a programme member committee and reviewer in several international journals, conferences and workshops. Stamatis serves in the technical advisory board of Internet Protocol for Smart Objects Alliance (IPSO), and the Permanent Stakeholder Group of the European Network and Information Security Agency (ENISA).

Prof. Jerker Delsing is the chaired Professor in Industrial Electronics Luleå University of Technology, Sweden since 1995. Present research profile is 'Embedded Internet Systems', EIS, which is an approach to Internet of Things, IoT and System of Systems, SoS. Here, applications are mainly found in industry automation. The general idea is that most sensors and actuator (low resources devices) will have communication capability using the Internet and the 'TCP/IP' protocol suite and be capable of ad hoc integration into a communication network and an application framework. Integration technologies are services, of object-oriented models. He has been the main supervisor of 17 students achieving the ph.D. degree and has also been the main supervisor of 24 students achieving the Licentiate degree. He is currently actively supervising 11 ph.D. students. His complete publication list is found at www.ltu.se/staff/j/jerker-1.11583. In summary, there are more than 25 journal papers and 70 conference papers in scientifically reviewed journals and conferences. He is currently the coordinator of Europe's largest automation project Arrowhead, www.arrowhead.eu.

Dr. Petr Stluka is Engineering Fellow in Honeywell Labs and lead of the Data-Centric Technology group since 2004, driving research efforts of the group in the fields of statistical modelling, predictive analytics, optimisation and decision support in application areas related to industrial process plants, buildings and homes. In his role, he is closely working with Honeywell technology and marketing leaders and contributing actively to strategic roadmaps. He received his M.Sc. in 1995 and ph.D. in 1998 from Prague Institute of Chemical Technology, both in the field of technical cybernetics. Petr Stluka has seven U.S. patents and has authored more than five other patent applications. He is author or co-author of more than 50 technical publications. His research interests include alarm management, event processing techniques and industrial energy efficiency.

Prof. Robert Harrison is Director of the Automation Systems Group in WMG at the University of Warwick. He has core expertise in the design and application of recon-figurable automation systems and related virtual engineering tools and methods. He is the author of over 150 peer-reviewed international journal and conference papers. Working very closely with the industry, Prof. Harrison has international experience in SOA and component-based systems engineering. He has been involved in many European and national projects including IMC-AESOP, SOCRADES and EPSRC Business Driven Automation, and his current research in the field of cyber-physical systems includes the EPSRC Knowledge Driven Configurable Manufactur-ing, ARTEMIS Arrowhead, and the TSB Direct Digital Deployment and Augmented Manufacturing Reality projects. Prof. Robert Harrison is Director of the Automa-tion Systems Group in WMG at the University of Warwick. He has core expertise in the design and application of reconfigurable automation systems and related vir-tual engineering tools and methods. He is the author of over 150 peer-reviewed international journal and conference papers. Working very closely with the industry, Prof. Harrison has international experience in SOA and component-based systems engineering. He has been involved in many European and national projects including IMC-AESOP, SOCRADES and EPSRC Business Driven Automation, and his cur-rent research in the field of cyber-physical systems include the EPSRC Knowledge Driven Configurable Manufacturing, ARTEMIS Arrowhead, and the TSB Direct Digital Deployment and Augmented Manufacturing Reality projects.

François Jammes was previously in charge of leading awarded collaborative projects such as ITEA SIRENA and SODA, and was the technical manager of the FP6 SOCRADES project. François Jammes was the director of the 'Web services' inter-nal Schneider Electric project, which was investigating and deploying the SOA con-cepts and Web services technology inside the group. He is a Schneider Edison group senior expert, reporting to the Buildings Business Unit, coordinating the relationship between this organisation and the other Schneider Business Units. He is involved in several Building Automation related European collaborative projects fromthe FP7 framework, and is also an expert for the European Commission in this application domain. He holds many European and international patents, and has published many related articles in IEEE journals and conferences.

Prof. Dr. Jose L. Martinez Lastra joined the Department of Production Engineering at the Tampere University of Technology (Finland) in 1999, and became Full Professor of Factory Automation in 2006. Prof. Lastra earned his advanced degrees (MS—with distinction and Dr. Tech.—with commendation) in Automation Engineering from the Tampere University of Technology. His undergraduate degree in Electrical Engineering is from the Universidad of Cantabria (Spain). His research interest is on applying ICT technologies to Factory Automation, with focus on manufacturing systems based on autonomous embedded networked production units. Previous to his current position, Prof. Lastra carried out research at the Department of Electrical and Energy Engineering (Universidad de Cantabria), the Mathematics Department (Tampere, Finland) and the Hydraulics and Automation Institute (Tampere, Finland). He was a visiting scholar at the Mechatronics Research Lab. of the Massachusetts Institute of Technology (Cambridge, MA). Prof. Lastra has authored over 200 scientific papers and holds a number of patents in the field of Industrial Automation. He has extensive experience in the industry as a consultant for the development of networked embedded control systems, including the first industrial implementations using Java-based embedded industrial controllers in the USA. Prof. Lastra joined the Department of Mechanical Engineering and Industrial Systems in January 2014 as the director of FAST-Lab, a research unit devoted to the seamless integration of human knowledge and intelligent machines/systems.

Index

A. W. Colombo et al. (eds.), *Industrial Cloud-Based Cyber-Physical Systems*,
DOI: 10.1007/978-3-319-05624-1, © Springer International Publishing Switzerland 2014

If you have any concerns about our products,
you can contact us on
ProductSafety@springernature.com

In case Publisher is established outside the EU,
the EU authorized representative is:
Springer Nature Customer Service Center GmbH
Europaplatz 3, 69115 Heidelberg, Germany

Printed by Libri Plureos GmbH
in Hamburg, Germany